Masonry for Sustainable Construction

Edited by Amjad Almusaed and Asaad Almssad

Published in London, United Kingdom

Masonry for Sustainable Construction
http://dx.doi.org/10.5772/intechopen.104068
Edited by Amjad Almusaed and Asaad Almssad

Contributors
Alexey N. Plotnikov, Viktor A. Ivanov, Boris V. Mikhailov, Tatyana G. Rytova, Olga S. Yakovleva, Mikhail Yu Ivanov, Natalia V. Ivanova, Manuel de Jesús Pellegrini Cervantes, Margarita Rodríguez Rodriguez, Susana Paola Arredondo Rea, Ramón Corral Higuera, Carlos Paulino Barrios Durstewitz, Alberto Muciño Vélez, Edrey Nassier Salgado Cruz, Eligio Alberto Orozco Mendoza, César Armando Guillén Guillén, Noemi Maldonado, Gerardo González del Solar, Pablo Martín, María Domizio, Praveen Kumar R., Balaji D.S., Navaneethakrishnan G., Amjad Almusaed, Asaad Almssad

First published in London, United Kingdom, 2023 by IntechOpen
IntechOpen is the global imprint of INTECHOPEN LIMITED, registered in England and Wales, registration number: 11086078, 5 Princes Gate Court, London, SW7 2QJ, United Kingdom

British Library Cataloguing-in-Publication Data
A catalogue record for this book is available from the British Library

Additional hard and PDF copies can be obtained from orders@intechopen.com

Masonry for Sustainable Construction
Edited by Amjad Almusaed and Asaad Almssad
p. cm.
Print ISBN 978-1-83768-125-9
Online ISBN 978-1-83768-126-6
eBook (PDF) ISBN 978-1-83768-127-3

Meet the editors

Prof. Amjad Almusaed has a Ph.D. in Architecture (Environmental Design) from Ion Mincu University, Bucharest, Romania. He completed postdoctoral research on sustainable and bioclimatic houses at the School of Architecture, Aarhus, Denmark, in 2004. His academic focus is on the role of sustainability in building design and city planning. He has done extensive study, research, and technical surveying in these fields. He is an active member of numerous international architectural organizations and has edited several books. His bibliography includes around 185 worldwide academic publications (articles, studies, books, and chapters) written in various languages.

Associate Professor Asaad Almssad has worked in industry, academia, and research for more than 30 years. He has held positions at Umea University, Sweden; Karlstad University, Sweden; and a variety of other European and non-European institutions. His studies concentrate on building materials, building structures, environmentally responsible construction, and the energy efficiency of building systems. He has more than sixty academic papers and numerous books to his credit. Currently, he is employed as a docent at Karlstad University, Sweden.

Contents

Preface

Bricks are some of the first building materials and masonry has been the base of construction for thousands of years. Bricks have been a staple in human construction for eons. Initially, people around the Nile, the Tiber, and the Euphrates utilized mud brick. Humankind employed the first burned bricks similarly to other finishing materials. The Great Wall of China, at 8851.8 kilometers in length, is the most extended artificial structure in the world. It was constructed in the third century BC and built with two brick walls and a stone base. Every two hundred meters along the wall, a watchtower was built. Six meters separate the brick walls at the bottom of the building, but they draw together for an extra half a meter of stability at the top. As a result, there is a strong layer of clay between the walls.

Bricks can be laid in irregular or front patterns and the bricklaying technique selected determines the long-term stability, dependability, and longevity of the building or structure. Because of its characteristics, brick serves as a structural and serial element. The aesthetic value of bricks has traditionally been recognized as one of their distinguishing features. Brick was used as a building material and a face material, giving rise to iconic pictures of cities worldwide; its aesthetic potential has yet to be fully realized. Moreover, this material's employment in building cladding was often determined by pragmatic concerns like cost-effectiveness and utility, leaving little room for innovative design. Brick veneer may be used in various configurations and styles to give structures their distinctive appearance.

Masonry is a type of construction in which the walls of a system serve as the primary structural element, and the masonry units are placed to provide structural support and load-carrying. Masonry is a time-tested and universal building material that has been utilized for decades or even centuries all over the globe. The material is eco-friendly since it can be recycled and reused when construction is complete. Masonry's resilience and durability are two of its most notable benefits. Masonry buildings are resilient and long-lasting, withstanding the wrath of Mother Nature even in the face of hurricanes, typhoons, and other extreme weather events. In addition, the long-term cost of maintaining a masonry structure can be decreased by its resistance to rot, vermin, and fire.

Bricks, concrete blocks, ceramic blocks, and other materials laid in a horizontal line make up what is meant by "one line" of masonry. A brick wall is a single-layer construction that consists of several courses of brickwork laid vertically. Masonry veneers are another name for single-layer, non-structural brick walls. By reinforcing and grouting a row of masonry to create lintels, sills, and ring beams, masonry walls may be made into more robust structural components. The search for new visual solutions and the development of new types of aesthetic options for architectural solutions is the issue currently facing wall ceramics, thanks to the expansion of civil engineering and developers' desire for individuality, expressiveness, and aesthetic diversity.

The term "masonry engineering" describes the practice of employing masonry materials, such as clay bricks, load-bearing clay hollow bricks, autoclaved lime-sand bricks, fly ash bricks, a wide variety of tiny and medium-sized blocks, and stone, in building construction. The construction materials might include bricks, stones, blocks, lightweight wall panels, and other materials. Masonry materials include stone or concrete blocks set in mortar. Mortar acts as a binding substance to bond the blocks together to support loads of daily use and other stresses placed on the building. It is important to consider both the masonry's location and the surrounding environment when choosing the type and grade of cement to use in the mortar. It is essential to verify the glue's strength and stability before using it. Batches used for inspection must be produced by the same company and have the same serial number. Masonry and stone masonry are the two most common kinds of masonry. Masonry executed using bricks is called "brick masonry." Cement masonry is the most cost-effective, whereas "clay work" utilizes clay to fill the seams between bricks in a wall. Masonry refers to construction with natural materials such as bricks or stone. Masonry's strength is measured by how well it bears the weight of the structure above it. Internal strains and deformations are produced when loads are applied to masonry. The strength of masonry varies depending on factors such as the kind of mortar and brick used, the size and form of the masonry elements, and the width and density of the mortar joints. Stability refers to masonry's capacity to remain in place when subjected to a horizontal load. Because of this quality, the maximum height of a masonry building is constrained by the material's thickness and the wind's force. Bricks of various materials (silicate, ceramic, face, refractory) are assessed for their heat conductivity. The thermal conductivity coefficients of refractory bricks are shown for a range of temperatures, from 20°C to 1700°C, allowing for a comparison of the bricks' thermal properties.

Masonry is the most common material used in the construction industry, which includes many other subfields. Rectangular pieces of burned clay, known as masonry, are used to construct buildings and walls. These days, eco-masonry may be crafted from a wide range of resources, such as recycled plastic bottles, clay, and so on, each of which has unique advantages.

This book takes a comprehensive look at contemporary masonry use in building. Masonry is interpreted in this book as a central theme of modern architecture and green buildings. As a substance, it ranks among the highest in human history for its value. Masonry is a sustainable building material used for centuries in construction. Its construction utilizes plentiful and recyclable resources, including clay, sand, and stone. Masonry is an adaptable building material because it may be used to construct either load-bearing or non-load-bearing walls. Several masonry constructions have survived for centuries. As a result, there will be less of a need for constant upkeep and repairs, which is great for the environment and for saving money. As masonry walls have a large thermal mass, they are effective in keeping the interior of a structure at a comfortable temperature. This is why it is cheaper to heat and cool brick structures than other types of structures and thus bricks are better for the environment. As masonry is permeable to water, it can lessen the likelihood of flooding and protect against damage caused by water. For this reason, masonry constructions are more resistant to hurricanes and tornadoes. Masonry's great fire resistance and inability to catch fire mean it may be used to keep flames from spreading and protect people inside buildings. Masonry may be recycled when a building's useful life is up, cutting down on landfill trash and saving valuable materials. Masonry is an environmentally friendly option that may help construct sturdy and long-lasting structures. Its application in building aids conservation, waste reduction, energy efficiency, and resistance to natural calamities.

Amjad Almusaed
Department of Construction Engineering and Lighting Science,
Jönköping University,
Jönköping, Sweden

Asaad Almssad
Faculty of Health, Natural Sciences, and Technology,
Department of Engineering and Chemical Sciences Construction Technology,
Karlstad University,
Karlstad, Sweden

Section 1

Masonry in the Construction Process

Chapter 1

Introductory Chapter: Bricks between the Historical Usage and Sustainable Building Concept

Amjad Almusaed and Asaad Almssad

1. Introduction to the thematic area

Bricks are the most frequent ceramic product and may be found in antique and modern structures. Together with stone and concrete, they are among the most used building materials [1]. Reading through history, we may see that throughout the Stone Age, cave dwellers erected structures for a variety of reasons out of fragments of rocks and boulders of various shapes; Menhirs, dolmens, and cromlechs have survived to this day—stone constructs used for religious purposes. Dwellings and fortresses were constructed from unhewn natural stone, the shards of which were piled on each other without any order. Brick is a common antique building material that has been used from ancient times, such as in Egypt's ancient dwellings, Rome's Colosseum, and many sections of China's Great Wall. One of the earliest construction materials is brick [2]. It was utilized in Mycenaean civilization, ancient Greece, Rome, Central America, and other ancient constructions. The oldest, used in prehistoric times, is dry masonry of irregularly shaped stones. Soil is a natural building material that can be seen everywhere, and it also has its unique style in traditional architecture. Like wood, stone has been the primary building material since ancient times. It has been used as a construction material since prehistoric times. The actual art of brickmaking may be observed in the great range of textures and surface treatments, which constitute a distinctive mark of each maker. Brick is now employed for resistant buildings, regular internal walls, interior or outside surface decorating, pavement, and even modern art installations. No other building feature provides as many opportunities for producing one-of-a-kind architectural effects. The most ancient type of bricks in the Western Hemisphere is a type of bricks as adobe [3]. Nicolas Durand et al. affirm that the precipitation of calcium carbonate is common in soils and regoliths, especially in soils of arid environments [4]. At the same time, calcareous porous clay can be found in dry regions worldwide but is mainly mined in Latin America, Mexico, and the southern United States are all included. Manzanilla et al. argue that the ancient Aztec tribe built the Pyramid of the Sun out of adobe in the fifteenth century and that it still stands today. In contrast to modern brick, ancient brick was square and flat (sides 30–60 cm, thickness 3–9 cm) [5]. There are generally two types of masonry: brick and stone masonry. Brick masonry: a type of masonry that uses bricks. However, masonry is further divided into "clay work," which uses clay to fill various joints with bricks to build walls, and "cement masonry," the cheapest type of masonry. Masonry: this is the art of building with bricks or stone. The ability of masonry to support the load imposed by the structural elements above it is called strength. Bricks are made of clay, shale, shale, and other materials that are crushed and pressed manually or mechanically. After drying, they are burned with an oxidizing flame at around 900 degrees Celsius [6]. Calcined bricks have varying hues depending on the kiln temperature and are extensively used worldwide as a simple, sturdy, and economical building material. Dimensional standards make it incredibly adaptable, and structures of all sizes may convey a nice texture practically throughout human history. The ability of masonry to maintain its position under horizontal load is called stability. Chen think that in the clay-rich river meadows of the Nile valley, people began making artificial bricks from clay much earlier using molds, which were solidified by mixing in chopped straw and camel dung. This technique is probably around 15,000 years old [7]. Almusaed and Almssad, wrote that a significant leap in development occurred with the "invention" of burnt brick. One is sure today that already around 4000 BC. Bids fired in Mesopotamia were known around 3000 B.C [8]. Bricks of various hues might be manufactured. According to the findings of excavations in Mesopotamia, Egypt, and other "cradles" of civilization, there were brick constructions long before Christ. Egypt's oldest burnt brick constructions are from the third and second centuries. They were constructed in Almusaed and Almssad, believe that the people who built such bricks simply took clay soil, soaked it, kneaded it, mixed it with different additions (straw, manure, chips, etc.), and dried the formed mass of butter in the sun after densely using the wooden forms with considerable force [9, 10]. After thousands of years of development, the brick has retained its virtues. Augustine Uche Chukwu Elinwa argues that when bricks were formed from silty soil with crushed straw in ancient times, and subsequently when low-melting clays and loams were mixed with sand, sawdust, ash, and other mineral components, the clay, water, and sand constituted the foundation of the "brick test" [11]. Brick has played a significant part in the history of world architecture. Construction methods employing bricks and stones are always changing. Brick's productivity has increased with time, making it the most preferred building material. Brick structures were frequently employed in civic and industrial constructions in the eighteenth century. Later, brick buildings were used to create warehouses, factories, and other infrastructure [12]. This chapter will discuss and analyze historic and modern buildings that use masonry as the main construction material.

2. Brick components and classification

A brick is an artificial stone that is formed into bars. After the fire, it is made of mineral components and develops stone-like qualities that effectively construct quite large structures, particularly country low-rise cottages [13]. Clay bricks are created from local resources, are inexpensive and long-lasting, and provide benefits such as fire protection, heat insulation, sound insulation, and moisture absorption. They are frequently employed in civil building projects. Initially, clay was utilized as a raw material to create bricks, but red clay with poor fertility was used in agriculture. This red clay brick is highly durable, and laterite bricks may have been found in many early human civilizations. However, because of the expansion of the industrial revolution and the emergence of environmental concerns in the twentieth century, people began to employ new raw materials to manufacture industrial bricks that did not require fire.

Silicate brick, according to the production procedure, controls around 10% of the market. This type of brick is made by pressing a combination of slaked lime and quartz sand into a hardened lime mortar in an autoclave. Silicate brick comprises 90% lime, 10% clay, and a tiny number of additives [14]. Bricks come in various hues, including blue, green, and purple. A specific number of pigments must be added to accomplish this. However, silicate brick's natural coloring is white or light gray. The advantages of such a brick are its inexpensive cost and capacity to give a range of shades. However, there are considerable limitations, including that it is a big brick, not robust enough, readily transfers heat, and is not water-resistant. This sort of brick is only utilized in the construction of partitions and walls. It is inferior to ceramic bricks in this regard [15, 16]. Brick made of ceramic. It's made by heating clay. Strength, fire resistance, good soundproofing, durability, the capacity to balance temperature variations, and not absorbing dangerous elements from the environment are all advantages of this type of brick. According to the application's features: Construction. Load-bearing walls and partitions are constructed using bricks, which are then encased, plastered, and painted. Facing (front, façade) (front, facade). It features large voids in the "body," contributing to the brick's improved thermal insulation capabilities. These bricks also offer great soundproofing capabilities [17]. A textured brick has a relief pattern on the surface. This type of brick is used to finish interiors and exteriors. Textured brick and shaped are two variations. Textured side faces have an irregular relief or a consistent geometric design. Shaped bricks occur in various forms: angular, beveled, semicircular, and notched. You may use this kind to create columns, vaults, and arches, as well as beautifully embellished cornices and windows. Furnace. This type of brick is constructed from refractory clay. It comes in a wide range of sizes and shapes [18]. Regarding the process of using the waste and broken bricks can be utilized as concrete aggregates. It produces lightweight, high-strength, hollow, and considerable blocks to address the disadvantages of typical clay bricks, such as small size, self-weight, and soil consumption [19]. Lime-sand bricks are created in proper proportions of lime and quartz sand, sand, or fine sandstone, which are pulverized, combined with water, semi-dry pressed, and autoclave cured [20]. Fly ash bricks are created by batching, molding, drying, and roasting using fly ash as the major raw material, combined with coal gangue powder or clay and other cementing materials, and may make full use of industrial waste left over while saving fuel.

3. A review on the historical use of bricks in high-value buildings

3.1 Brick work in old cities

Bricks, also known as masonries, are cuboid construction elements burned from the earth and were known as smelting tiles in ancient times in Mesopotamia and Egypt. Bricks, often gray or red brick in color, can be used as ornamental patches on walls or as accurate load-bearing structures and are the primary building materials for local residences, churches, and institutions in earthquake-prone locations. Brick represents the first product specially made for construction, where the first bricks—mud bricks—were clay bars, which were molded in wooden frames and dried naturally in a hot climate. The oldest find, indicating the demand for brick construction, dates to the eighth millennium BC [21] (see **Figure 1**). These bricks were discovered in the Middle East, and structural research revealed that they were formed of clay, mud, and resin; the bars were molded by hand and dried in the sun. Mud bricks were manufactured in ancient Egypt and Mesopotamia from clay silt gathered from a few rivers. According to Pinakin Dhandhukia et al., tiny pebbles and chopped straw were added

Figure 1.
Historical usage of bricks in important buildings.

to the raw material to boost the strength and prevent shrinking [22]. The Egyptians built the wall with dried bricks and a liquid clay mortar, while the Assyrians employed newly molded material that was subsequently bonded into a monolith. Because raw brick was so strong, it was used to build large structures. It was frequently paired with stone or charred brick. By the way, the latter was chosen not just for its greater strength but also for its natural longevity. Burnt bricks were used to construct palaces, temples, and other places of worship. According to Han, Lim Chung, et al., clay burnt bricks are commonly utilized as filler between structural frames in the building industry. Builders employ this approach because it is well-known, simple to construct, and does not necessitate the use of skilled personnel [23]. Mesopotamian cities were constructed with hewn-dried bricks and lime and gypsum mortar. Many localities have employed clay mortar [24]. Pozzolana, a hydraulic binder, was used for construction mortar in ancient Rome. It was built with hewn stones and ceramic brick masonry; residential homes and other domestic constructions were built with various additions of dried, unburnt clay bricks. The petrology of these samples can disclose a plethora of information, not only about the processes occurring in these lime-ceramic combinations but also about the nature of the pottery employed as an ingredient [25].

According to Kadim Hasoon, the ancient population of Mesopotamia in Iraq (from the Stone Age, 150,000–8000 BC) is one of the earliest civilizations known. After that, they were Old Stone Age (Paleolithic) (150,000–12,000 BC): Around 100,000 years ago, humans lived in caves in northern Iraq and made their instruments out of stones. In Sulaymaniyah, the oldest caves are Zerzi cave and Hazar Murd cave. 1 The entrance of one of the most famous caverns in the Zagros Mountains, "Shanider," is 8 meters high and 25 meters wide. From the interior, it measures 40 meters long and 53 meters broad. 2 Clay brick masonry is one of humanity's oldest and most lasting construction methods [21]. Brick burning was invented by the ancient Sumerians thousands of years ago. Images from South Iraq demonstrate how bricks were collected and constructions were built with them. Furthermore, the difference between that and the construction location is not considerable. The ancient Sumerians utilized a triangle to check the correctness of the wall building, according to Rossi C. and Almusaed, and the bricks were worn on the yokes [19, 26].

The underlying notion of building construction has remained virtually unchanged [27]. In China, for example, there are two primary forms of masonry used as

Figure 2.
Brickwork from China the (Qin dynasty) and Mesopotamia (Sumerian antic).

construction materials: sintered bricks (clay bricks) and non-sintered bricks (lime-sand bricks and fly ash bricks). Clay bricks, according to Amin Al-Fakih et al., are formed of clay or shale, with coal gangue powder as the significant essential ingredient. To manufacture it, it must go through mud treatment, molding, drying, and roasting operations [28]. "Bricks and tiles from Qin and Han"! This is because ancient China's Qin and Han dynasties were periods of outstanding architectural adornment. The technology, manufacturing size, quality, and decorative variants of brickmaking have all advanced significantly [29]. Opus maxima is the last kind of Roman stonework (layered masonry). This idea means that the wall was constructed of various materials, often alternating layers of brick and stone (see **Figure 2**).

The use of brick in Mesopotamian and Ancient Roman architecture is highly significant, particularly in the province of Ancient Italy ruled by the Etruscans. They constructed their temples from raw bricks and embellished them with terracotta features. Brick in such structures is already taking on a more recognizable elongated form for us. Bricks are unusual among Mesopotamian artifacts because "textual evidence attests that they integrate standards of length, area, volume, capacity, and weight-a very uncommon combination in the history of pre-modern metrology." Different laying methods were utilized in Greek dwellings for a thousand-year BC (according to Homer) and were made from raw dried clay bricks. For many years, burnt brick was the dominant building material in Roman and Byzantine architecture. Brick work was completed on lime mortar with crushed brick chips added. Occasionally, stone rows would alternate [30]. Europe gratefully absorbed the experience of peoples and millennia. On the territory of Germany, brick gave its name to a whole style of architecture—brick Gothic dominated here during the twelfth to sixteenth centuries. Bricks were first acknowledged in Europe in the fourth century. According to Dmitriy Yakovlev et al., it was used to build castle walls, temples, towers, and furnaces. Plinths were thin, heavy slabs of varied sizes used in the eleventh and twelfth centuries. In the fifteenth century, a brick in the shape of a bar arose, like the present one [31]. Many European towns have historic brick cathedrals that serve as examples of European architecture. The material's reach was enlarged in the Middle Ages; it was also employed as a decorative and expressive means: craftsman constructed patterned brickwork from curved (shaped) glazed bricks [32]. As an example, let us cite the European patterned architecture of the sixteenth to seventeenth centuries. In the middle of the nineteenth century, production was nevertheless mechanized: a belt press was used for molding, and a ring kiln was used for firing. In the same years, the dimensions of the brick stabilized (the format, however, with some rounding, has been preserved to this day). Most brick buildings in Europe were constructed from crudely processed natural stone in the tenth to twelfth centuries; brick was only employed in a few facilities to level the masonry. With the advancement of brick manufacturing technology, the use of lime mortar for clay laying is becoming increasingly frequent. According to Maldonado, Noemi, and others, the masonry of the time was marked by a significant thickness of the seam: the masonry mortar was created from broken bricks and was noted by exceptional strength. Until the sixteenth century, as brick construction technology advanced, the ways of installing hewn natural stone continued to improve everywhere [33]. By the early seventeenth century, brick had firmly established itself as the primary material for the construction of residential structures. The size of the brick was practically current for the period but varied from maker to manufacturer. Even back then, builders and producers began to consider the necessity for a single-size standard for bricks. According to Joseph and Tretsiakova-McNally, it was conceivable to make a uniform raw brick in the second half of the nineteenth century: a set of standards for state-owned companies arose, limiting the size of an already fired product rather than a raw brick [34]. Emmitt and Prins consider that in the 1930s, it was also considered necessary to fill the vertical joints

Figure 3.
Shows the St. Pancras Renaissance London Hotel & St. Pancras railway station designed by arch. George Gilbert Scott.

of the masonry with mortar, which guaranteed wind proofs: this work was detailed and required a lot of time and effort [35]. In the United Kingdom, the neo-Gothic style was evident in Victorian architecture, including railway stations, residential structures, government buildings, public institutions, and cathedrals. Isabelle Cases confirms that the distinctiveness of Victorian architecture originates from the fact that it still accounts for a significant amount of Great Britain's current constructed history. Many well-known public facilities, churches, and stations, as well as entire streets and neighborhoods and even specific urban planning decisions, were built during the Victorian era [36]. One of the most striking examples of this direction is the Midland Grand Hotel (today the St. Pancras Renaissance London Hotel), which also serves as the facade of St. Pancras railway station. Architect George Gilbert Scott designed it with the participation of engineer William Barlow (see **Figure 3**). The hotel's main façade is made of unbaked brick and covered with chimneys and crenelated gables, thus contrasting with the station's minimalist landing stage [37].

3.2 Brick used in early modern architecture

With the resurgence of cities around the end of the tenth century, brick apartment structures on two or three stories with workshops and stores below began to be erected. Patterned masonry was created, and it frequently employed brick with a figured surface that was coated in a strong, lustrous finish. True, it was expensive, and only wealthy clients—monarchs, monks, and great feudal lords—could afford it. Brick has been employed as a fundamental aspect of architecture by architects across the world and throughout history. Great exponents such as Antoni Gaud, Puigi Cadafalch, and Doménechi Montaner developed a place with its own individuality in the Mediterranean: a land of bricks [38]. Today, Spanish contemporaries such as Rafael Moneo, "brick magicians" such as Uruguayan Eladio Dieste, and recent Pritzker laureates such as Paulo Mendes da Rocha and Glenn Murcutt employ brick as a building technology. Brick's ability to act as a structure and an enclosure is one of the "keys" to understanding it as a modern material. For most of architectural history, these two jobs were carried out by masonry: the building of a supporting structure that has steel or concrete reinforcement. Architects and theorists held different viewpoints on the brick in the twentieth century as a result of the division of duties made feasible by the end of the nineteenth century.

Nevertheless, use of the brick persisted throughout the twentieth century, and it was integrated into virtually every architectural style. Nevertheless, numerous paths and methods of valorizing the "traditional" brick in connection to modern materials such as steel and concrete may be recognized. A first tendency is the reinterpretation of brick in the spirit of modernist ideas, even if it is employed in load-bearing walls, with a preference for basic, relatively plain surfaces; Daniel Burnham's design for the Monadnock Building in Chicago 1889 exemplifies this direction as shown in **Figure 4**.

From the Wainwright Building, a second, more detailed direction may be observed (Louis Sullivan, St. Louis, 1890) [39] because the bricks have been built such that they might propose a hotel timetable from behind. Sullivan's approach, viewed as an example of "sincerity in design," had a tremendous effect on early twentieth-century architecture (see **Figure 5**).

A third option was influenced by the design of nineteenth century enterprises with load-bearing brick walls. The constructions in this category are distinguished by large flat brick surfaces, functional volume distribution, and specific load-bearing components, such as "pillars" made of wood or cast iron. Their architecture combines

Figure 4.
The Monadnock Building in Chicago is depicted above (1889) P. Berlage's Amsterdam Exchange is seen in the lower picture (1903).

Hotel Imperial (Tokyo, 1906

Buffalo, 1904

Figure 5.
Important building from early modernism.

classic and contemporary elements. In the context of "expressionist" architecture of the early twentieth century. Between the wars, a youthful school of European Modernist architects experimented with new spatial notions based on Cartesian orthogonality, intersecting planes, and abstract cubic forms using brick. When the bearing brick was employed in Modern architecture, it was usually coated with a smooth coating of plaster. Architects in the United States appeared less engaged in the ideological conflict between emerging Modernist aesthetics and the usage of traditional materials. Brick was employed as the primary building material (American Standard Building, 1923), but it was also used in conjunction with steel (Crysler Building, 1930). Following WWII, the use of brick—both in load-bearing walls and in enclosures—was "revitalized" because of a new interest in raw materials that may be represented in a very direct and forceful manner [40].

The load-bearing brick, whether wall, pilaster, or arch, had a little influence on twentieth-century architecture. It is critical to understand how architects employed bricks throughout history, a technical study of brick manufacture and masonry, and an essay on architectural and cultural history. The brick approach was used by modernist architects such as Louis Kahn, Alvar Aalto, and Renzo Piano to provide rich content. In addition to cathedrals, country residences, temples, and mosques are shown. Louis Khan for example built his own profound, mystical, and everlasting work style over 50 years, steadily ascending to the ranks of architectural masters. He built more than 10 masterpieces in the last 15 years of his life, which established the size and standard of American architecture and dramatically affected the evolution of American architecture. He didn't leave many structures, but practically everyone has become a "Mecca Holy Land" for architectural aficionados. They are breathtakingly gorgeous, with clean geometric designs, raw and honest materials, and lovely light-created spaces. His career was not prolific, but his best works were unique and displayed their beauty in shocking new ways. "You asked Brick: 'Brick, what do you want to be?'" he wrote on his use of brick in his work. I adore arch coupons,' Brick told you. 'Look, I also want arch coupons, but arches are costly; I can purchase them if you put a concrete lintel above the opening,' you remarked to Brick. Then you ask, 'Brick, what do you think?' "I adore arches," says Brick." Louis Kahn's. Dan and Dianne Chrzanowski, originally from Cleveland, had the opportunity to live out their love after more than 20 years of studying Frank Lloyd Wright's work by becoming the owners of the Wright-designed Dobkins House, a house in which the architect employed brick to express its message [41].

Brick was most noticeable in non-load-bearing enclosures in the postwar period. Industrial structures of the nineteenth century with load-bearing brick walls Modernist works accepted the evident brick closure as a compromise; the historical resonance of its texture is completely absorbed into the Modernist language. "Post-modernism" makes exaggerated or humorous allusions to the tradition represented by the brick in the second half of the twentieth century; architecture is less bound by modernist formal aesthetics and more openly related to the historical past [42]. In Europe, different postmodernism arose, as opposed to the United States: the "essence" of bricks—weight, resistance to compression, solidity—were recognized and employed in an old component synthesis framework.

3.3 Brick in post-modern and contemporary architecture

Brick manufacturing frequently uses a lot of natural resources. Reusing industrial wastes to create sustainable bricks is a recent development in research and application that aims to lower the carbon footprint of the brick production sector [43]. Nevertheless, brick remains a popular and commonly used building material. But to develop a chic and distinctive place, interior designers and decorators are constantly rediscovering him. Brick is now frequently utilized as a decorative element in the shape of a brick wall, which may make an interior stylish and pleasant. It should be mentioned that not everyone likes indoor brickwork. Some people think brick in the inside is too rough; others think brick in the interior looks dreary.

On the other hand, most enthusiasts of mixing solidity and unfettered creativity are content to employ a brick wall inside their structure. This innovative method is appropriate for practically any home design. Nonetheless, the brick wall is frequently the focal point of the living area. Surprisingly, it will serve as a good foundation for current technologies (e.g., a plasma TV). Furthermore, a fireplace will seem quite natural next to a brick wall. Because brick is a raw material, it is safe to use in the bedroom and nursery. You may create the sense of a medieval cell in the bedroom by decorating the wall at the head of the bed with a brick. Because of its heat resistance, a brick wall will be helpful in the interior of the buildings. Brickwork, for example, may be utilized in the kitchen, where it symbolizes a new value in modern design, and it is also fitting because stoves and chimneys are typically installed there. A brick wall is usually the focal point of the space (see **Figure 6**). Preliminary heat treatment is applied to this construction material. As a result, the brick isn't terrified of mold [44]. Such a wall will look excellent and preserve its original aspect even after many years.

The use of a brick wall in the inside of a building is one of the most popular decorating ideas. Most importantly, such a decorative feature is appropriate for both rural and loft designs. The growing popularity of old industrial buildings aided in the growth of brick interiors, and the brick wall formed the foundation for the loft design. Brick is frequently utilized to accent one wall in modern homes that must be noticed. A brick wall remains a dynamic part of the interior even if it is not red brick, but rather yellow, white, or gray. Facing brick is one of the most popular materials for completing unique and unusual interiors. It might be matte, aged artificially, or glazed. To adhere the brick to the wall, use the same glue as you would when working with imitation stone. The seams must be sealed with a tone-matching tile grout. Because facing brick is lighter and thinner than natural brick, it conserves useable area and is ideal for facing tiny rooms and thin walls. There are several methods for decorating rooms with bricks. Still, when designing your interior, choosing a strategy that suits the room's original data, the eventual aim, and your financial capabilities is preferable. Traditional load-bearing masonry building technologies encountered

Figure 6.
Shows modern Kitchen brickwork.

new structural and environmental control systems and new functional, spatial, and aesthetic criteria over the twentieth century [45]. Nevertheless, the brick has continued to be used mostly for closings, despite competition from more contemporary and technologically complex materials, demonstrating that its "non-structural" features are still valued: cost-effectiveness, flexibility, durability, impact resistance, and, last but not least, appearance Furthermore, in the context of current concerns for "sustainable" architecture, the "ecological" qualities of ceramics—as a "healthy" material with a low environmental impact—as well as the ecological control capacity of the massive wall, bring the single-layer masonry back to the forefront, but based on products with improved performance.

4. Bricks between traditional using and environment

The designer should recognize the need to return to healthy construction materials. Construction materials must be made or processed in the near neighborhood, requiring the shortest possible transit pathways. When producing building materials, as little energy as feasible is consumed. The rawer materials that are changed, the more electricity is required. Brick is a common raw material in Scandinavia, and its durability and low cost make it one of the most extensively used and recognizable raw materials in Scandinavian architecture. Modern architects frequently include conventional construction materials and modern discoveries and technology into their buildings. On the other hand, materials such as brick, wood, and stone have long been utilized to produce both basic and distinctive structures. Brick has distinct qualities that contribute to its continued use as one of the most frequent building materials [46]. A human views ordinary brickwork differently from a concrete wall. A concrete wall conveys monotony, but a brick wall conveys vitality. Throughout history, brick has been used to bring beauty and utility to places by providing protection for internal spaces while allowing natural light and circulation to penetrate. Today, architects and builders have many options when it comes to building cladding materials, and no matter which option they choose, they must find a balance between their architectural vision and the needs of their customers. Durability, minimal maintenance, and long-term sustainability have always been essential assessment factors, in addition to the aesthetics and qualities of the chosen materials.

The brick wall perfectly withstands such loads. This material is also suitable for constructing walls and interior partitions. Various hinged structures and objects are well attached to such walls with the help of dowels and nails. A brick wall is characterized by high thermal inertia; it cools slowly and warms up slowly. This means that the air temperature does not change much in houses with such walls during the day. Walls below the level of waterproofing can only be made of clay bricks, well-fired. And the corrugated surface of the brick is an excellent basis for applying plaster and bonding various mortars together. To create complex architectural forms, the construction of buildings with columns and vaults, and solid bricks are also used. In this case, it is necessary to construct supporting columns for the burials of towers and domes. The strength and durability of brick make it the best material for this purpose. Therefore, both earlier and now temples and churches are built of brick. Brick is also an ideal building material for building basement walls. A feature of basements is the frequent contact of their borders with soil and groundwater. Therefore, the walls are subject to constantly changing temperatures, moisture, etc. Material properties such as strength, resistance to moisture, and other natural influences make it ideal for building basements of buildings.

Brick new construction Ceramic and silicate bricks are the two varieties employed in constructing new brick constructions. The more costly ceramic brick performs admirably, being incredibly resilient to heat and moisture. Although silicate brick is substantially less expensive, it cannot withstand high humidity and extreme temperature changes. Currently, hollow bricks are the most common type. Their low weight and superior thermal insulation are their key benefits. The brick material, which is extensively used across the world, is not only recyclable and extremely resistant to natural dangers such as as fire, storm, and moisture, but it also has a high degree of ease of use in size, form, color, and texture, as well as a cheap cost and high adaptability. Environmentally friendly building materials are given less attention. It is difficult to achieve environmentally responsible construction without employing ecologically friendly goods and resources. Brickwork is starting to be used for larger projects and public buildings and complexes; new materials and forms are appearing and being created, and construction site robots are faster and more precise than humans. All types of bricks, in comparison with concrete, are environmentally friendly and contain a minimum number of additives. Moreover, all additives in the material's structure are non-chemical, with a low carbon footprint. According to many specialists, brick buildings are the most comfortable—they have high rates of heat resistance, hygroscopicity, and sound absorption, and environmental friendliness is also at its best.

5. Conclusions

Since the beginning of civilization, humans have been responsible for producing and using a component that has been essential to developing cultures and civilizations. There are times when we tend to forget that man cannot carry out his ambitions without the most fundamental tools. When we think about the house of our dreams, the one in which we will raise our children and enjoy the company of our grandchildren, we think about the fundamental and unbreakable labor of the brick, which settles into our lives and becomes a testament to our life. When we look at this house, we think about the fundamental and unbreakable labor of the brick. Clay bricks are currently facing technological challenges and are uncompetitive when compared with materials such as concrete, where rising environmental concerns about the accumulation of unmanaged wastes from agricultural or industrial productions have made these promising candidates for incorporation into building materials to improve their performance. Over the long history of brick existence, the primary method of creating burnt ceramic bricks has remained unchanged: clay extraction and preparation, preparation of a mixture, shaping, and, lastly, fixing the resulting shape and structure. Since then, nothing has changed in regard to the fundamental ideas of building structures. Adobe brick was once widely used in the West. It was made from porous lime clay with the addition of minerals such as quartz, resin, and other materials, and the resulting "bricks" were then sun-dried. In this chapter, we looked at bricks as a primary building material, where they are used in modern architecture, their significance in historical and contemporary architectural styles, how they are used in modern architecture, and various projects of different kinds, such as private and public facilities in various climes and other places.

Author details

Amjad Almusaed[1*] and Asaad Almssad[2]

1 Department of Construction Engineering and Lighting Science, Jonkoping University, Sweden

2 Faculty of Health, Science and Technology, Karlstad University, Sweden

*Address all correspondence to: amjad.al-musaed@ju.se

References

[1] Crespo-López L, Cultrone G. Improvement in the petrophysical properties of solid bricks by adding household glass waste. Journal of Building Engineering. 2022;**59**:105039

[2] Bel-Anzué P, Elert K. Changes in traditional building materials: The case of gypsum in Northern Spain. Archaeological and Anthropological Sciences. 2021;**13**:177

[3] Henn T, Nagy DU, Pál RW. Adobe bricks can help identify historic weed flora—A case study from south-western Hungary. Plant Ecology and Diversity. 2016;**9**(1):113-125

[4] Nicolas DH, Monger C, Canti MG. 9—Calcium Carbonate Features, Interpretation of Micromorphological Features of Soils and Regoliths. UK. 2010. pp. 149-194

[5] Manzanilla L, López C, Freter A. Dating results from excavations in quarry tunnels behind the pyramid of the sun at teotihuacan. Ancient Mesoamerica. 1996;7(2):245-266

[6] Kadir AA, Sarani NA. An overview of wastes recycling in fired clay bricks. International Journal of Integrated Engineering. 2012;**4**(2):53-69

[7] Chen M. Harappa Civilization. In: China and the World in the Liangzhu Era: Liangzhu Civilization. Singapore: Springer; 2022

[8] Almusaed A, Almssad A, Najar K. An innovative school design based on a biophilic approach using the appreciative inquiry model: Case Study Scandinavia. Advances in Civil Engineering. 2022;**2022**:42-54

[9] Almusaed A, Almssad A. Building materials in eco-energy houses from Iraq and Iran. Case Studies in Construction Materials Journal. 2015;**2015**:42-54

[10] Homod RZ, Almssad A, Almusaed A, et al. Effect of different building envelope materials on thermal comfort and air-conditioning energy savings: A case study in Basra city, Iraq. Journal of Energy Storage. 2021;**34**:101975

[11] Elinwa AU. Effect of addition of sawdust ash to clay bricks. Civil Engineering and Environmental Systems. 2006;**23**(4):263-270

[12] Mohammed Ali Bahobail. The mud additives and their effect on thermal conductivity of adobe bricks. JES. Journal of Engineering Sciences. 2012;**40**(1):21-34

[13] Lum C et al. Potential use of brick waste as alternate concrete-making materials: A review. Journal of Cleaner Production. 2018;**195**:226-239

[14] Hamburg J, Lorenzon M. Before Meeting the Greeks: Kutaisi Influence in Late Bronze and Early Iron Age Colchian Settlements. Journal of Field Archaeology. 2022;**47**(1):13-31

[15] Almusaed A, Alasadi A, Almssad A. A research on the biophilic concept upon school's design from hot climate: A Case Study from Iraq. Advances in Materials Science and Engineering. 2022;**2022**:12 pages. Article ID 7994999. DOI: 10.1155/2022/7994999

[16] Adamo N, Al-Ansari N. Babylon in a New Era: The Chaldean and Achaemenid Empires (330-612 BC). Journal of Earth Sciences and Geotechnical Engineering. 2020;**10**(3):87-111

[17] Almusaed A, Yitmen I, Almssad A, Homod RZ. Environmental profile on

building material passports for hot climates. Sustainability, Sustainability. 2020;**12**(9):3720. DOI: 10.3390/su12093720

[18] CultroneItziar G, Carmen A. Casado, Anna Arizzi, Sawdust recycling in the production of lightweight bricks: How the amount of additive and the firing temperature influence the physical properties of the bricks. Construction and Building Materials. 2020;**235**:117436

[19] Rossi C. On measuring ancient Egyptian architecture. The Journal of Egyptian Archaeology. 2020;**106**(1-2):229-238

[20] Almusaed A, Almssad A. Lessons from the World Sustainable Housing (Past Experiences, Current Trends, and Future Strategies), In: Sustainable Housing. London, United Kingdom: IntechOpen; 2021 [Online]. DOI: 10.5772/intechopen.100533. Available from: https://www.intechopen.com/chapters/79055

[21] Hnaihen KH. The appearance of bricks in ancient Mesopotamia. Athens Journal of History. 2020;**6**:73-96

[22] Dhandhukia P, Goswami D, Thakorb P, Thakker JN. Soil property apotheosis to corral the finest compressive strength of unbaked adobe bricks. Construction and Building Materials. 2013;**48**:948-953

[23] Han LC et al. Use of compressed earth bricks/blocks in load-bearing masonry structural systems: A review. Materials Science Forum. 2020;**997**:9-19

[24] Selvaraj T, Devadas P, Perumal JL, Zabaniotou A, Ganesapillai M. A comprehensive review of the potential of stepwells as sustainable water management structures. Watermark. 2022;**14**:2665

[25] Siddall R. From kitchen to bathhouse: The use of waste ceramics as pozzolanic additives in Roman mortars. In: Ringbom Å, Hohlfelder RL, editors. Building Roma aeterna: current research on Roman mortar and concrete. San Francisco, California: The Finnish Society of Sciences and Letters. 2011

[26] Almusaed A. Biophilic and Bioclimatic Architecture, Analytical Therapy for the Next Generation of Passive Sustainable Architecture. London Limited, London, UK: Springer-Verlag; 2011. pp. 123-130

[27] Al-Fakih A, Mohammed BS, Liew MS, Nikbakht E. Incorporation of waste materials in the manufacture of masonry bricks: An update review. Journal of Building Engineering. 2019;**21**:37-54

[28] Nadali D. Esarhaddon's glazed bricks from Nimrud: The Egyptian campaign depicted. Iraq. 2006;**68**:109-119

[29] Jia Q, Zhou Y, Shen YN. Qin Brick and Han Tile in Chinese Ancient Construction—Aesthetic Analysis of the Ancient Chinese Building Materials in Qin and Han Dynasties. AMM. 2013;**357-360**:282-284

[30] Karydis N. Early Byzantine Vaulted Construction in the Western Coastal Plains and River Valleys of Asia Minor. British Archaeological Reports International Series. 2011;**2246**:143-149

[31] Yakovlev D, Trushnikova A, Antipov I. The cross-cultural interaction in the Baltic region in the fifteenth century: The vaults of the Faceted Palace in Novgorod the Great and Brick Gothic architecture. Journal of Baltic Studies. 2020;**51**(4):553-568

[32] Almssad A, Almusaed A, Homod RZ. Masonry in the context of sustainable

buildings: A review of the brick role in architecture. Sustainability. 2022;**14**:14734

[33] Maldonado NG, Martín P, Solar GG, Domizio M. Historic Masonry. In: Turcanu-Carutiu D, editor. Heritage. London: IntechOpen; 2019

[34] Joseph P, Tretsiakova-McNally S. Sustainable non-metallic building materials. Sustainability. 2010;**2**(2):400-427

[35] Emmitt S, Prins M, editors. Proceedings of the CIB W096 Architectural Management: New Directions in Architectural Management. Lyngby, Denmark: Technical University of Denmark; 2005

[36] Boerefijn W. About the Ideal Layout of the City Street in the Twelfth to Sixteenth Centuries: The Myth of the Renaissance in Town Building. Journal of Urban History. 2016;**42**(5):938-952

[37] Stamp G, Scott GG. In Search of the Byzantine: George Gilbert Scott's Diary of an Architectural Tour in France in 1862. Architectural History. 2003;**46**:189-228

[38] Collins GR. Antonio Gaudi: Structure and form. Perspecta. 1963;**8**:63-90

[39] Condit CW. Sullivan's Skyscrapers as the expression of nineteenth century technology. Technology and Culture. 1959;**1**:78-93

[40] Goldhagen SW. Something to talk about: Modernism, discourse, style. Journal of the Society of Architectural Historians. 2005;**64**(2):144-167

[41] Siry JM. Frank Lloyd Wright's Innovative Approach to Environmental Control in His Buildings for the S. C. Johnson Company. Construction History. 2013;**28**(1):141-164

[42] Introductory chapter: overview of a competent sustainable building. In: Almusaed A, Almssad A, editors. Sustainable buildings - interaction between a holistic conceptual act and materials properties [Internet]. London: IntechOpen; 2018 [cited 2022 Dec 21]. DOI: 10.5772/intechopen.77176. Available from: https://www.intechopen.com/chapters/61634

[43] Chin WQ, Lee YH, Amran M, Fediuk R, Vatin N, Kueh ABH, et al. A sustainable reuse of agro-industrial wastes into green cement bricks. Materials. 2022;**15**:1713

[44] Faiz Ahmed ATM et al. Hemp as a potential raw material toward a sustainable world: A review. Heliyon. 2022;**8**:e08753

[45] Kirschke P, Sietko D. The function and potential of innovative reinforced concrete prefabrication technologies in achieving residential construction goals in Germany and Poland. Buildings. 2021;**11**:533

[46] Al-Fakih A, Mohammed BS, Ehsan MS-L. Nikbakht, Incorporation of waste materials in the manufacture of masonry bricks: An updated review. Journal of Building Engineering. 2019;**21**:37-54

Chapter 2

Recycled Conductive Mortar

Manuel de Jesús Pellegrini Cervantes,
Margarita Rodríguez Rodriguez, Susana Paola Arredondo Rea,
Ramón Corral Higuera and Carlos Paulino Barrios Durstewitz

Abstract

Due to the urgent need to care for the environment, the use of recycled materials is necessary. The creation of multifunctional materials with content of recycled materials presents an alternative to reduce the use of natural resources. This is through the addition of recycled fine aggregate, product of industrial waste in its manufacture, such as graphite powder (GP) and carbon fiber (CF), turning it into conductive recycled mortar (CRM). The sustainability of this new material brings great ecological benefits, such as the reduction in the use of fine aggregates, which are naturally present in rivers, and also, lower production of construction waste sent to landfills. In this research, an evaluation of the effect of the addition of carbon fiber and graphite powder on wet, dry and hardened electrical properties, electrical percolation in dry state, and flowability of the mixture of recycled conductive mortar in a wet state-based on cement——fine aggregate from waste blocks——graphite powder was carried out. The results obtained showed the effect of the addition of GP and CF to the mortar mix, mainly the reduction of its flowability, caused by the physical interaction between the recycled sand or recycled fine aggregate RFA and the carbon fiber CF, as well as the graphite powder GP.

Keywords: sustainability, multifunctional material, electrical conductivity, workability, fluidity

1. Introduction

Cement-based mortars have a very low electrical conductivity (EC), in dry state, they are considered insulating materials, and in wet state, they are considered semiconductors; thanks to the ionic conductivity provided by the pore solution in its cementitious matrix. To improve the EC of cement-based mortars, the addition of conductive materials in the form of powder and/or fiber, such as CF, GP and carboxymethylcellulose (CMC). These materials are considered suitable conductive additions, since they have high EC and mechanical and physical properties that help to transform the cement-based mortar into a sustainable multifunctional material, by giving it mechanical and electrical properties of significant utility, not only for structural use but also for electrical and/or electrochemical use. In the case of GP and CF, their physical properties, such as length, particle size, average diameter, percentage of addition and dispersion in the cementitious matrix, determine the effect produced as an addition in mortars. The dispersion and concentration of CF considerably affect the air content of the mortar mixture, as it alters the mechanical and elastic properties of the material. The inhomogeneous dispersion of the conductive material in the cement matrix, either GP and/or CF, generates a negative effect on the mechanical and electrical properties of the material; compressive strength, tensile strength, flexural strength, conductivity and electrical percolation. A homogeneous dispersion of the CF in the cementitious matrix allows reaching electrical percolation limits with lower percentages than those of GP additions, given the high aspect ratio of the CF, obtaining multifunctional materials of high EC with low volumetric percentages of CF, a complex situation in the use of GP. For fine aggregate, it is possible to use fine recycled aggregates RFA obtained from concrete, produced by mechanical crushing, composed of natural fine aggregate coated with hardened mortar or paste, which results in lower mechanical strength, different setting times and higher water absorption. Its use is an action that contributes to the sustainability of construction materials [1–19].

2. Methodology

2.1 Materials

For the manufacture of conductive recycled mortars, composite Portland cement (CPC) 30R type I, RFA recycled fine aggregate, GP loresco SC-3 graphite powder, CF carbon fiber with a length of 10 ± 1 mm, carboxymethylcellulose and distilled water were used. The properties of the materials are shown in **Table 1**.

The carbon fiber (CF) used for the manufacture of CRM specimens was obtained from industrial waste products, it was carefully selected and cut for use in the CRM mixtures. The diameter of the CF was constant, of continuous morphology, smooth and free of defects. A photograph of its cutting and selection is shown in **Figure 1** item a). **Figure 1** item b) shows a micrography of carbon fiber CF.

The concrete used to obtain the RFA recycled fine aggregate was waste material from the quality control laboratory; the concrete was crushed using a jaw crusher. Subsequently, the retained material was selected between mesh no. 4 and no. 50, to guarantee the absence of fines resulting from crushing, and cement dust, avoiding the production of a mortar mixture with high water demands. The particle size of the RFA was determined according to ASTM C 136 [20], ASTM C 33 [21], and ASTM C 125 [22] standards. **Figure 2**, item a) shows the crushing process and item b) the material after the crushing process, ready for the particle size test.

Carboxymethylcellulose CMC was also used as an emulsifier in the CRM mixtures, as this material has physical properties that allow it to act as an adherent between GP, CF, RFA and CPC particles.

The particle size distribution of the cement used is shown in **Figure 3** item a. Regarding the coarse particles, the CPO has particles exceeding 100 μm, and the fine

Material	Average Diameter (μm)	Superficial Area (m^2/gr)	Specific Density
Carbon Fiber CF	7.000	0.227	1.760
Graphite Powder GP	204.000	2.290	1.850

Table 1.
Properties of materials used in the fabrication of CRM.

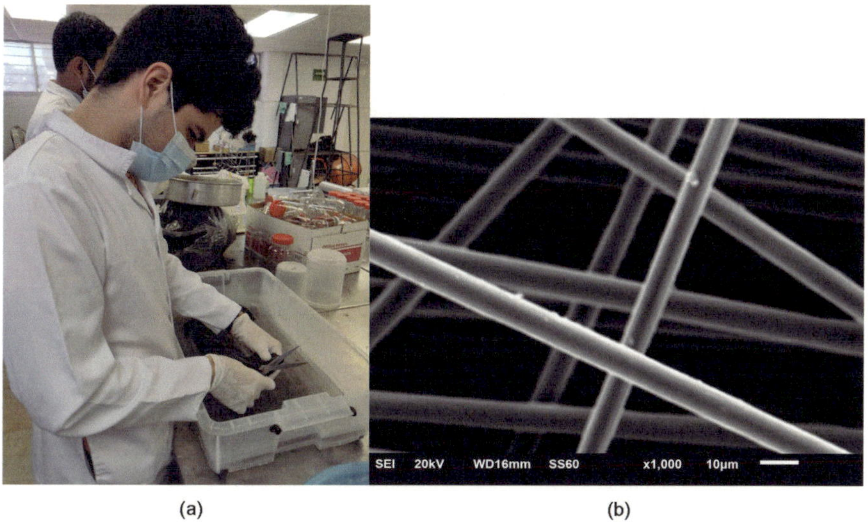

Figure 1.
a) Cut and selection of carbon fiber CF. b) Micrography of carbon fiber CF.

Figure 2.
a) Concrete crushing process. b) Recycled fine aggregate RFA after crushing process.

particles have sizes up to 1.0 μm. According to SEM shown in **Figure 3** item b, the particles that make up the cement are irregular, but with smoother edges with sizes from 1.5 to 50.0 μm. The particle size distribution of the cementitious materials, and their physical and chemical properties define the microstructure of the hardened mortar, some of these properties are shown in **Table 2**.

2.2 Conductive mortar specimens

The electrical conductivity (EC) was determined in specimens of CRM recycled conductive mortar with 1, 3, 7, and 28 days of curing in distilled water, specimens

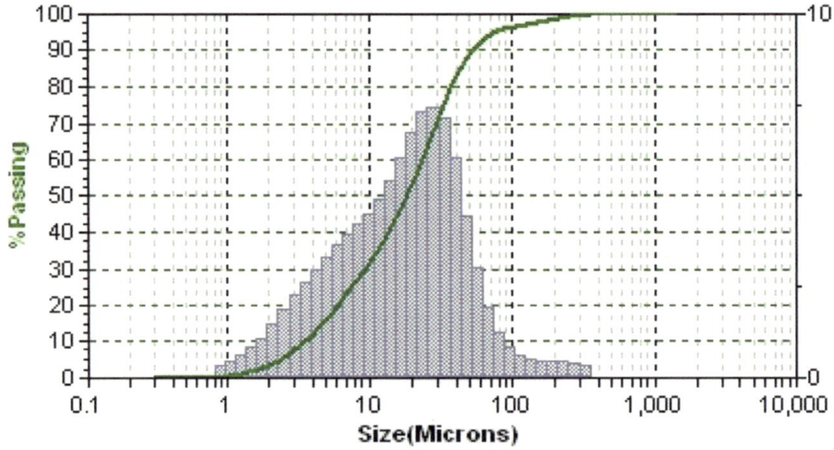

(a)

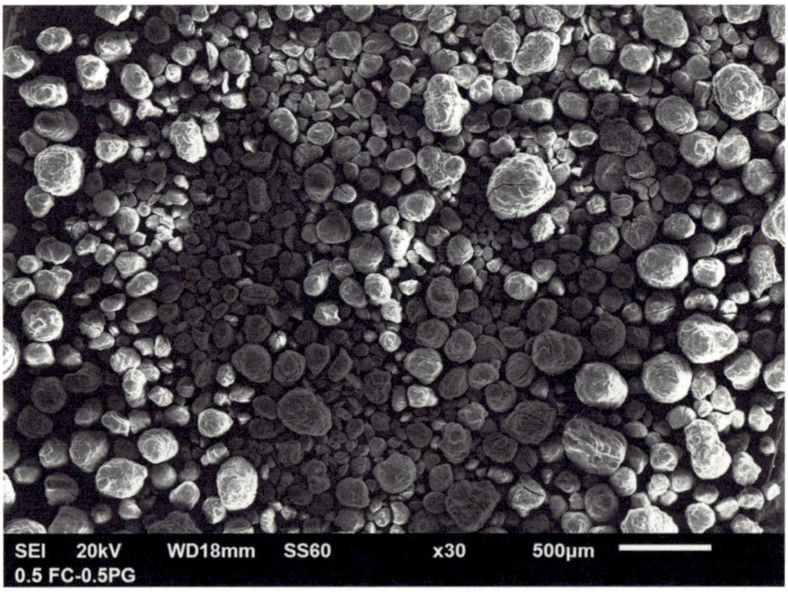

(b)

Figure 3.
a) Cement particle size distribution. b) Cement SEM.

Property	CPO
Mass density (kg/m^3)	2975
Superficial Area (m^2/g)	19.26
Average particle size (Φm)	27.61

Table 2.
Particle size distribution of cementitious materials.

were manufactured in the shape of a quadrangular prism of *40 mm x 40 mm x 160 mm*. The material ratios for the mixtures were: sand/cement = 1.00, water/cement = 0.60, and CF = 0.1, 0.2, 0.3, 0.4, 0.5, 1.0, and 1.5% with respect to the weight of cement. Two types of specimens were manufactured, with and without GP, in both cases, using the CF percentages mentioned above. **Table 3** shows the dosages used in the mixtures.

The process for the fabrication of CRM with CF and GP established by **Table 3** was carried out using the procedure described in ASTM C 305–14 [22] with the following variants:

1. For the case of CRM type M-GP-CF, the RFA was manually mixed with the GP until a homogeneous material in visual appearance was achieved, prior to the start of the mixture manufacturing.

2. The CF was dispersed with the total mixing water and the CMC in ultrasound for 30 minutes.

3. The CF was placed with the total water in a mixing vessel, adding the total cement and mixing at a slow speed of 140 ± 5 r/min for 30 s.

4. The total amount of recycled sand and GP, if any, was added slowly for 30 s while mixing at a slow speed of 140 ± 5 r/min.

5. Mixed for 30 s at an average speed of 285 ± 10 r/min.

6. The mixer was stopped, remaining covered at rest for 90 sec. In the first 15 s the walls of the container were scraped quickly, using a stainless-steel spatula.

7. It was mixed for 60 s at an average speed of 285 ± 10 r/min.

Figure 4 shows the industrial mixer used during the process.

Mortar	CF percentage in relation to cement weight							Graphite/ Cement Ratio
M-CF	0.10	0.20	0.30	0.40	0.50	1.00	1.50	0.00
M-CF-GP	0.10	0.20	0.30	0.40	0.50	1.00	1.50	1.00

Table 3.
Dosages for CRM mixtures.

Figure 4.
Industrial mixer used during the process.

2.3 Electrical conductivity determination

After 24 hours had elapsed since the mixtures were placed in the molds, the electrical conductivity (EC) was determined for the specimens made with the mixtures according to the dosages in **Table 3** for the ages of 1, 3, 7, and 28 days. The electrical resistivity was measured using 2 methods: four-point method with miller 400A resistivity meter (4 PM-RM) and four-point direct current method (4 PM-DC) according to [23, 24]. The experimental arrangement is shown in **Figure 5**, the resistivity was determined from Eq. (1)) and the electrical conductivity (EC) with Eq. (2)).
where:

$$\rho = Fm * 2 * \pi * a * R \tag{1}$$

$$\sigma = \frac{1}{\rho} \tag{2}$$

- ρ = Resistivity (Ω.cm).
- σ = Conductivity (S/cm)
- Fm = Geometric factor involving the length of the specimen (L)
- and the separation between the electrodes

- a = Separation between the electrodes (cm)
- R = Electrical resistance (Ω)

Fm was determined based on the L/a ratio and methodology proposed by Morris et al., 1996 [25] and Garzón et al. 2014 [23]. Fm corresponds to 0.1547 for the used dimensions of the CRM specimens and the electrode spacing.

3. Environmental problems

It is recognized that the use of natural fine aggregate, commonly extracted from rivers, has a negative impact on the environment since the demand for this material grows over time and its presence in the environment becomes scarce. In addition, the concrete waste deposited in landfills, a product of construction waste, is highly polluting and affects aquatic ecosystems.

The question is to define the potential that exists in replacing natural fine aggregate with recycled fine aggregate, obtained directly from the crushing of concrete-based construction waste. It is important to determine the possible benefits of using RFA recycled fine aggregate, since this, if proven as a viable alternative to the use of natural fine aggregate, implies a significant reduction in the exploitation of ecosystems and natural resources, as well as the preservation of the environment and, of course, sustainability.

The research "Percolation of a conductive recycled mortar", considers the sustainable development goals (SDGs) presented by the United Nations (UN). In particular, they are the following:

Objective 9. Build resilient infrastructure, promote inclusive and sustainable industrialization and foster innovation.

Objective 12. Ensure sustainable consumption and production patterns.

Objective 13. Take urgent action on climate change through education and public awareness.

Objective 17. Strengthen and revitalize the global partnership for sustainable development.

4. Theoretical basis of the research

- ASTM C 136-06. Sieve particle size analysis of fine and coarse aggregates.
- ASTM C-33. Establishes limits on harmful or noxious substances in fine concrete aggregate.
- ASTM C 125-00. Verification of mixing equipment, information required for mix design.
- ASTM C 305-14. For mechanical mixing of hydraulic concrete, cement pastes, and mortars of plastic consistency.
- ASTM C1437-13. Standard test method for flowability of hydraulic cement-based mortar.

5. Analysis and results

5.1 Dispersion of the carbonaceous material

The PCC with weight ratio CT/C = 0.50 and MS with a ratio of 0.10 with respect to the weight of the cement, which was manufactured with a water/cement ratio of 0.60, was checked for dispersion of the carbonaceous material by scanning electron microscopy (**Figure 5**). With the mechanical mixing procedure, uniform dispersion of the carbonaceous material was achieved.

Figure 6 shows the PCC, it can be seen that the crushed coke particles are uniformly dispersed in the cement matrix, presenting a spherical geometry and with a particle size less than or equal to 0.25 mm.

The addition of the DM in the mixture favors the dispersion of the carbonaceous material since, in mixtures made without this addition, the material is segregated at the ends of the specimens.

The segregation of the material without the addition of CMC does not allow a uniform contact between the grains of the material and impairs the electrical conductivity.

The dispersion of the carbonaceous material favors the electrical conductivity of the PCC, since the conductivity is due to the contact of the carbonaceous material grains, even in its dry state.

5.2 Granulometry of recycled fine aggregate

The granulometry of the crushed aggregates to be used in the preparation of the mixtures was verified by taking representative samples to determine their granulometry, according to **Table 4**. The fineness modulus is 2.77 for fine aggregate.

The particle sizes are not within the particle size limits specified by the ASTM C33 standard, as shown in **Figure 6**.

5.3 Electrical conductivity

The simultaneous use of GP and CF produces a synergistic effect on the properties of EC when incorporated in pastes and mortars [23]. **Figure 4** shows the

Figure 5.

Experimental arrangement for determination of EC electrical conductivity.

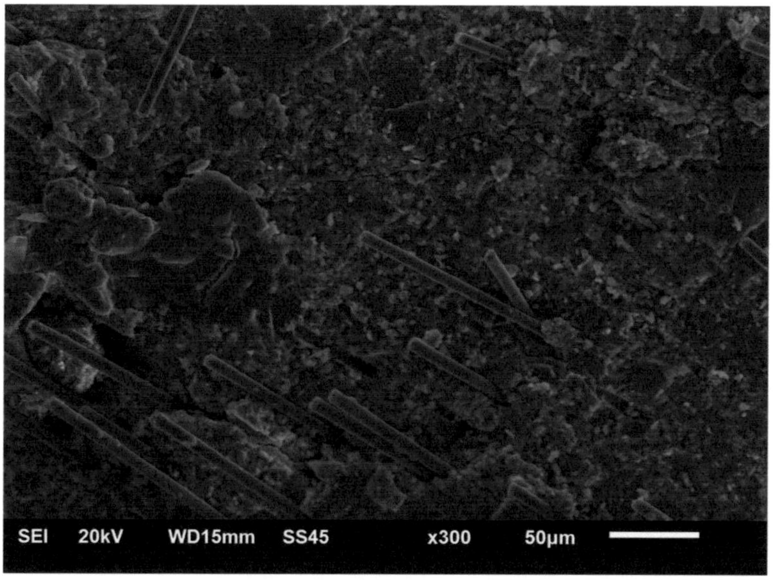

Figure 6.
Micrography of conductive cement paste weight ratio of CT/C = 0.50, 50X.

Mesh	Aperture (mm)	Retained weight (g)	Partially retained %	Accumulated retained %	Passing %
No. 4	4.76	0.00	0.00	0.00	100.00
No. 8	3	140.00	30.16	30.16	69.84
No. 16	1	151.20	32.57	62.73	37.27
No. 30	0.63	103.70	22.34	85.07	14.93
No. 50	0.5	64.80	13.96	99.03	0.97
Suma		459.70		276.99	
Fineness modulus = 2.77					

Table 4.
Granulometry of fine aggregates.

behavior of the EC in the wet state for different % CF with respect to the weight of cement; increasing the % CF produces increases in EC for all curing ages, in approximately the same proportions. Therefore, it is confirmed that the EC does not depend on the age of the RCMs, which makes it impossible to define an electrical percolation threshold, due to the contribution of the ionic conductivity of the pore solution to the EC of the RCMs, as in the case of M-CF mixtures. Similarly, **Figure 5** shows the EC in dry state for different % of CF for 28 days of age. As the % CF increases, there are increases in the EC, the most notorious being the one presented when using 30% of CF, so it was determined that this value is the percolation threshold, since the values for subsequent EC did not change much with respect to the 30% of EC (**Figures 7–9**).

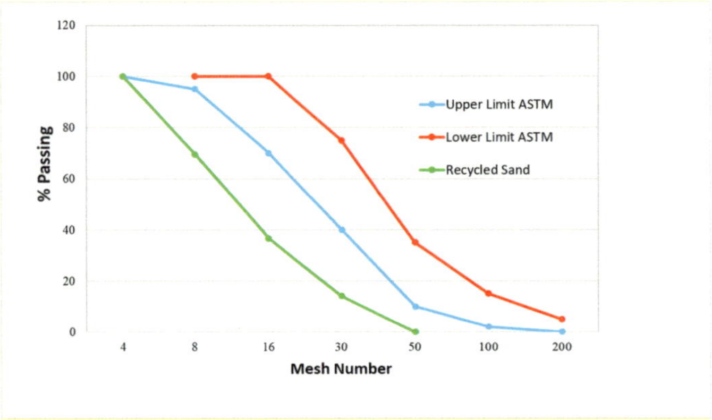

Figure 7.
Granulometry of fine aggregates according to ASTM C33.

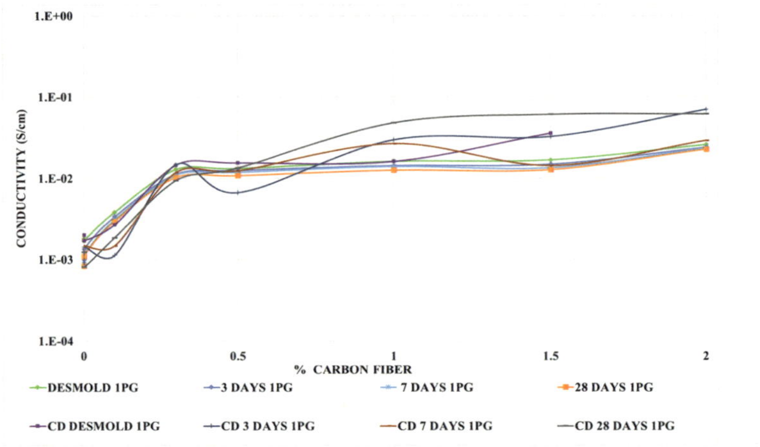

Figure 8.
Conductivity in CRM type M-GP-CF in wet state.

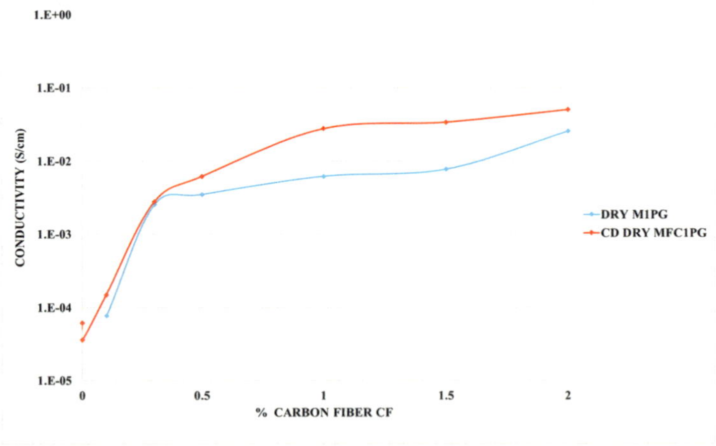

Figure 9.
Conductivity in CRM type M-GP-CF in dry state, 28 days of age.

6. Conclusions

Cement-based CRMs—recycled fine aggregate—graphite powder with added CF are an alternative as a multifunctional material, as well as sustainable, because they promote the reuse of recycled materials, which is beneficial for the environment and the construction market. These CRMs present a rapid decrease in flowability when the percentage of CF increases due to the physical interaction between the CF and the RFA in the wet state. For percentages above 1.0% of CF, the mixture with GP is no longer workable, with a tendency to inhomogeneous dispersion of CF and high air contents; in the case of CRM with the absence of GP, this situation occurs after 2.0% of CF. The addition of CF in CRM reduces the fluidity of the mixtures due to the opposition generated by its interaction with RFA and GP, in addition to the viscosity contributed by the Carboxymethylcellulose CMC, in its case. The electrical percolation threshold for CRM with GP content was estimated at 0.30% CF, below the case of no GP with 0.45% CF. This is because the increases in EC without GP are governed by the contact between the CF and the conductive pathways they form, while with GP the EC is defined by the contact between the CF and GP simultaneously, forming conductive pathways with higher EC performance.

Abbreviations

ASTM	American society for testing and materials
CF	carbon fiber
CMC	carboxymethylcellulose
CPC	composite portland cement
CRM	conductive recycled mortar
EC	electrical conductivity
GP	graphite powder
SEM	scanning electron microscope
4 PM-DC	four-point direct current method
4 PM-RM	four-point method with miller 400A resistivity meter

Author details

Manuel de Jesús Pellegrini Cervantes*, Margarita Rodríguez Rodriguez, Susana Paola Arredondo Rea, Ramón Corral Higuera and Carlos Paulino Barrios Durstewitz
Mochis School of Engineering, Autonomous University of Sinaloa, Mexico

*Address all correspondence to: manuel.pellegrini@uas.edu.mx

References

[1] Chen B, Wu K, Yao W. Conductivity of carbon fiber reinforced cement-based composites. Cement and Concrete Composites. 2004;**26**(4):291-297

[2] Shu X, Graham RK, Huang B, Burdette EG. Hybrid effects of carbon fibers on mechanical properties of Portland cement mortar. Materials and Design. 2015;**65**(1):1222-1228

[3] Chung DDL. Introduction to Carbon Composites. Carbon Composites. second ed. Butterworth-Heinemann; 2017. pp. 88-160

[4] Li M, Yang Y, Liu M, Guo X, Zhou S. Hybrid effect of calcium carbonate whisker and carbon fiber on the mechanical properties and microstructure of oil well cement. Construction and Building Materials. 2015;**93**:995-1002

[5] Han B, Zhang L, Zhang C. Reinforcement effect and mechanism of carbon fibers to mechanical. Construction and Building Materials. 2016;**125**:479-489

[6] Wang T, Xu J, Bai E, Luo X, Chen H, Liu G, et al. Study on the effects of carbon fibers and carbon nanofibers on electrical conductivity of concrete. Earth and Environmental Science. 2019;**267**:1-9

[7] Gao J, Wang Z, Zhang T, Zhou L. Dispersion of carbon fibers in cement-based composites with different mixing methods. Construction and Building Materials. 2017;**134**:220-227

[8] Jing X. Wu Yao y Ruiqing Wang, nonlinear conduction in carbon fiber reinforced cement mortar. Cement & Concrete Composites. 2011;**33**:444-448

[9] Wanga H, Gao X, Liu J. Miao Ren y Anxun Lu, multi-functional properties of carbon nanofiber reinforced reactive powder concrete. Construction and Building Materials. 2018;**187**:699-707

[10] Felekoglu B, Tosun-Felekoglu K, Ranade R, Zhang Q, Li VC. Influence of matrix flowability, fiber mixing procedure, and curing conditions on the mechanical performance of HTPP-ECC. Composites Part B: Engineering. 2014;**60**(4):359-370

[11] Yoo D-Y. Ilhwan you y Seung-Jung lee, electrical and piezoresistive sensing capacities of cement paste with multi-walled carbon nanotubes. Archives of Civil and Mechanical Engineering. 2018;**18**:371-384

[12] Wei J, Zhang Q, Zhao L. Lei Hao y Zhengbo Nie, effect of moisture on the thermoelectric properties in expanded graphite/carbon fiber cement composites. Ceramics International. 2017;**43**(14):10763-10769

[13] Wang D. Qiang Wang y Zongxian Huang, investigation on the poor fluidity of electrically conductive cement-graphite paste: Experiment and simulation. Materials and Design. 2019;**169**:107679

[14] Zhang B, Tian Y, Jin X, Tommy Y. Lo y Hongzhi cui, thermal and mechanical properties of expanded graphite/paraffin gypsum-based composite material reinforced by carbon fiber. Materials. 2018;**11**(11):2205

[15] Pellegrini-Cervantes MJ, Barrios-Durstewitz CP, Núñez-Jaquez RE, Baldenebro-Lopez FJ, Corral-Higuera R, Arredondo-Rea SP, et al. Performance of carbon fiber added to anodes of conductive cement-graphite pastes used in electrochemical chloride extraction in concretes. Carbon Letters. 2018;**26**:18-24

[16] Katz A, Kulisch D. Performance of mortars containing recycled fine aggregate from construction and demolition waste. Materials and Structures. 2017;**50**(4):199

[17] Yacoub A. Assia Djerbi y teddy fen-Chong, the effect of the drying temperature on water porosity and gas permeability of recycled sand mortar. Construction and Building Materials. 2019;**214**:677-684

[18] Zhang Y, Luo W, Wang J, Wang Y. Yaqin Xu y Jianzhuang Xiao, a review of life cycle assessment of recycled aggregate concrete. Construction and Building Materials. 2019;**209**:115-125

[19] Norma ASTM C 136-06. Standard Test Method for Sieve Analysis of Fine and Coarse Aggregates. USA: ASTM International; 2006

[20] Norma ASTM C 33-99a. Standard Specification for Concrete Aggregates. USA: ASTM International; 1999

[21] Norma ASTM C 125-00. Standard Terminology Relating to Concrete and Concrete Aggregates. USA: ASTM International; 2000

[22] ASTM International. Standard Practice for Mechanical Mixing of Hydraulic Cement Pastes and Mortars of Plastic Consistency. West Conshohocken, PA, USA: ASTM C 305-14. ASTM International; 2014

[23] Morris W, Moreno EI, Sagués AA. Practical evaluation of resistivity od concrete in test cylinders using a Wenner array probe. Cement and Concrete Research. 1996;**26**(12):1779-1787

[24] Garzon AJ, Sanchez J, Andrade C, Rebolledo N, Menéndez E, Fullea J. Modification of four point method to measure the concrete electricalresistivity in presence of reinforcing bars. Cement & Concrete Composites. 2014;**53**:249-257

[25] ASTM International. Standard Test Method for Flow of Hydraulic Cement Mortar. West Conshohocken, PA, USA: ASTM C1437-13. ASTM International; 2013

Chapter 3

Modeling of Structural Masonry

Gerardo González del Solar, María Domizio, Pablo Martín and Noemi Maldonado

Abstract

Masonry is a composite material, and its behavior shows that its weaknesses lie in the minimum resistance of its components and the characteristics of the interfaces between them. Ceramic brick masonry has technological characteristics that make it suitable for housing and building functions. The bricks, of reduced dimensions and joined with mortars of variable characteristics, have the advantage of adapting to almost all construction projects considering the influence of the environment on their service life. The investigation of the structural behavior of masonry has had very significant advances in the laboratory during the last mid-century, which has allowed numerical modeling of the behavior of the material and validation of failure modes under seismic actions. The behavior of heritage masonry with thick walls differs greatly from simple masonry using conventional techniques and materials. These differences in behavior have only been confirmed through numerical simulation contrasted with experimental research. This chapter presents the numerical modeling used for simple and confined masonry with reinforced concrete and for very thick heritage masonry, using the finite element method validated with full-scale laboratory experiences.

Keywords: masonry modeling, earthquake, thickness, simulation, FEM

1. Introduction

Masonry is a material composed of natural or manufactured units, generally joined with mortar, which constitute the inventory of existing constructions in the world from the Egyptian civilization to the present. Architecturally, there is a wide spectrum of uses in walls, arches, vaults, domes, beams, and columns that exhibit simplicity and elegance, but the analysis of its structural behavior is complex in heritage masonry. The most investigated construction techniques correspond to the Greek and Roman buildings that have remained standing to this day. In Africa and Asia, the earliest masonry was made of stone or earth. In America, stone masonry has been used in the pre-Columbian era, earth or adobe masonry in colonial times, and fired ceramic masonry from the end of the eighteenth century to the present with different variants of arrangement and combination of layers of different materials (**Figure 1**) [1–3].

The conservation of heritage buildings requires knowledge to guide preservation strategies. Materials degrade over time when they are in contact with the environment, and this is a natural and unavoidable process, and it is necessary to determine

Figure 1.
Examples of masonry in America (a) Machu Picchu (Perú) XV century, (b) Inca bridge and thermal hotel (Argentina) 1925, (c) chapel in Cuzco (Perú) 1598, (d) chapel in Córdoba (Argentina 1668), (e) king hovel's (Argentina-Chile) 1773, (f) interior Giol winery (Argentina), (g) Giol winery facade (Argentina) 1896, and (h) fader house museum (Argentina) 1889.

the rate of degradation, which is necessary data to estimate the service life of the construction in relation to safety and/or functionality. The presence of moisture that can come from the ground, rain, or faulty drainage services damage the masonry, especially ancient masonry. Interventions with new materials have often increased moisture problems in masonry [4, 5].

The effect of earthquakes has been devastating in unreinforced masonry constructions, but it has made it possible to summarize the problem of the damage generated by these horizontal vibratory actions [6, 7]. Earthen masonry has shown its lack of earthquake-resistant capacity over time in the different continents, even for low-magnitude earthquakes.

Repair and replacement materials are required to be chemically and mechanically compatible with the original materials. Environmental conditions require control of porosity and permeability to water vapor. There are records of damage to historic masonry due to failure to assess the compatibility of the repair material in terms of strength, density, and stiffness of the original material [8].

According to the available materials, the climatic conditions, and the functional requirements, a variety of types of masonry can be found, with traditional practices and local technologies that vary according to the different countries.

Masonry can be classified according to: the material (adobe, stone, brick, block), its location (country or city), its use (residential or public), the structural system (simple, confined, reinforced), time of construction (ancient, before the First World War, between the First and Second World Wars, after the Second World War, after the adoption of unified international codes) [2]. An important aspect to evaluate in the behavior is if the masonry resists the permanent and seismic loads generated by its own mass and the contributions of floors or ceilings supported on it [9].

Since the 1970s, studies and research have been carried out applying computational mechanics to achieve mathematical models that simulate the structural response of historical masonry due to its weakness under seismic actions. The complexity and uncertainty of the geometry of old buildings and the nonlinear behavior of masonry require an important contribution of computational analysis techniques.

2. Masonry modeling

Numerical modeling of masonry requires computational models that can capture the different failure modes and that are sufficiently accurate and simple to implement. There are several modeling techniques. The technique to be used starts from the desired level of precision and simplicity [10].

Finite Element Models (FEMs) and Structural Element Models (SEMs) represent the behavior of masonry at different scales, so predictions may differ significantly depending on the model chosen.

According to the current codes, the use of more complex models is not recommended due to the need for a great experience of the designer, high sensitivity of the parameters used, dispersion of the predictions, and the need for a better interpretation of the postprocessing to obtain results applicable [11].

Masonry exhibits different directional properties due to the influence of mortar joints acting as planes of weakness. Depending on the orientation of the joints and the directions of the stresses and, on the other hand, the level of normal stress applied, failure can occur only at the joints (bed joint sliding shear mode) or simultaneously at the joints together and bricks (bed joint slip shear mode or diagonal stress cracking). The significant number of influencing factors, such as the dimension and anisotropy of the bricks, the thickness of the joint and the arrangement of the bed and head joints, the material properties of both the brick and the mortar, and the quality of the coat site construction makes simulation of brick masonry extremely complex.

Masonry can be modeled as single-phase, bi-phase, or tri-phase material [11].

As a single-phase material, all the elements that make up the masonry: brick masonry + mortar + unit-mortar interface make up a homogeneous, isotropic or anisotropic continuum (**Figure 2a**), without differentiation of elements. This procedure is often preferred for the analysis of large masonry structures (macro-model), but it is not suitable for detailed stress analysis of a small panel, due to the difficulty of capturing all its failure mechanisms.

As a biphasic material, the expanded units are represented as continuous elements while the mortar joints and the unit-mortar interface are grouped into discontinuous elements (**Figure 2b**). This procedure is applicable to a wider range of structures because it reduces computational processing times (simplified micro-model).

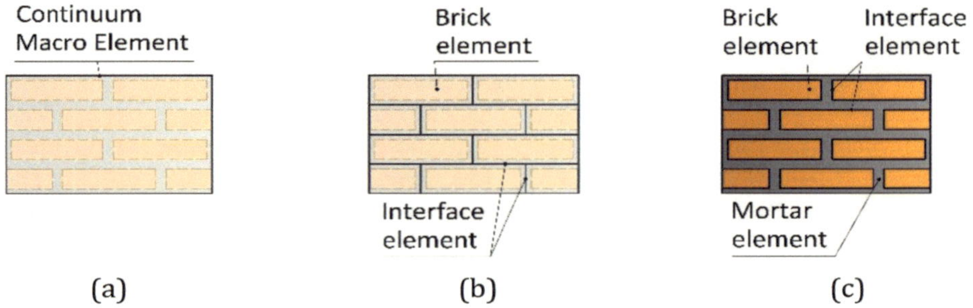

Figure 2.
Masonry modeling strategies: (a) macro-model, (b) simplified micro-model, and (c) detailed micro-model [11].

As a three-phase material, the masonry and the mortar joint are represented as continuous elements, while the unit-mortar interface is represented as a discontinuous element (**Figure 2c**). With this degree of mesh refinement, more accurate results can be obtained, but the availability of more powerful computational means is required (detailed micro-model), limiting its application to laboratory samples or structural details.

The boundary conditions that exist at the interface between the masonry and the surrounding frame in confined masonry have been modeled with springs or interface elements. The function of these elements is to represent the interaction between deformable structures, along surfaces where separations and sliding can occur.

Computationally, different commercial finite element packages have been developed for nonlinear, two-dimensional or three-dimensional static and dynamic analysis. All these software (ABAQUS, ADINA, ANSYS, ATHENA, and DIANA among the

Types of masonry	Details	Geometric Models	Macroelements	Blocks	Continuum Models
Unreinforced masonry	Single material	Static theorems Cinematic theorems	Equivalent frame Equivalent spring	Interface Contacts Texturized continuum Limit analysis Extended finite elements	Direct Homogenization Multiple scale
	Composed of various materials				
Confined masonry	Integrated masonry		Box behavior		
	Separate masonry		Equivalent structure		
Reinforced masonry	Vertical and horizontal bars included in masonry		Equivalent structure		

Table 1.
Masonry models.

most applied) incorporate libraries that include different finite element models and robust resolution strategies for different types of materials and load states, especially for brittle materials such as masonry and concrete. The use of these packages is required to resolve complex interfacial boundary conditions [6].

Table 1 presents a summary of the masonry models classified according to the structural system [2, 6, 9].

2.1 Modeling of unreinforced masonry

The classification proposed by D'Altri et al. [6] summarizes in a very complete way the investigations of the last 60 years in four types of generalized models for unreinforced masonry structures: models based on geometry, on macroelements, on blocks and as continuous material.

2.1.1 Geometry-based models

The masonry is modeled as a rigid body defined by the geometry of the structure. Structural equilibrium and collapse are studied through solutions based on limit analysis, which can be based on static or kinematic theorems (**Figure 3**).

Applications of the static theorem of limit analysis in real masonry structures are based on simple static schemes [13] suitable for the investigation of equilibrium states in arches, vaults, and domes, bounded between two extreme equilibrium conditions for static safety. In fact, if compression-only forces lines can be found within the confines of a vault, then the vault will remain in compression. Also, if the solution is within the middle third of the section, any stress (and thus any joints) will be present in the section.

There are different computational developments for the equilibrium analysis of masonry vaults, an analogy between the equilibrium of arches and hanging ropes (funicular model), the analysis of the thrust network that conceives vaults as membranes without tension. Few of these solutions have been able to incorporate the horizontal actions generated by earthquakes.

Kinematic theorems have been used in the last decades for an agile evaluation of masonry buildings. The Italian code has adopted the kinematic limit analysis approach, based on the decomposition into rigid blocks based on the failure mechanisms observed during earthquakes [14].

More advanced computational static theorem-based approaches have been developed to accurately assess the collapse multiplier and collapse mechanism of masonry structures. However, these approaches cannot provide the deformation capacity of a masonry structure, although they are very powerful to quickly and effectively assess the main vulnerabilities of a masonry building.

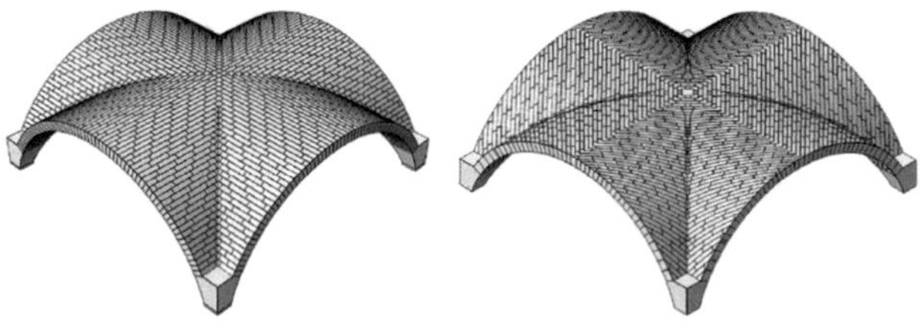

Figure 3.
Geometry-based model [12].

2.1.2 Macroelement models

When we refer of macroelement models, we mention a structure modeled in structural components at real or panel scale (1:1), considering a phenomenological effect or a nonlinear constitutive response, with the main structural elements being pillars and parapets. Observations of earthquake damage have shown that damage is concentrated on pillars and sills or lintels. With this structural idealization, the analysis of the global seismic response of the masonry construction is carried out.

Macroelement models are generally based on the assumption that any local failure mode activation, primarily associated with the out-of-plane twist response of masonry walls, is avoided. The seismic response is directly related to the shear capacity in the plane of the walls and to the load transfer due to the existence of diaphragms.

Both static and incremental dynamic global analyses are usually performed on 3D models, to consider load transfer between load-bearing walls due to horizontal action.

Columns are the vertical resistant elements that support vertical or horizontal loads. In contrast, spandrels or lintels are the horizontal portions of the structure between two vertically aligned openings, which couple the response of adjoining columns when loaded horizontally. Although the identification of masonry pillars and parapets can be easy in the case of masonry facades with regularly distributed openings, it becomes more complex when there are irregularly arranged openings, not being applied to very complex geometries.

Macroelement models are the most widespread modeling strategies used for the seismic evaluation of masonry structures due to their ease of computational application (also in 3D structures), together with the simple and fast definition of the model and the mechanical properties. The most used macroelement models correspond to equivalent beams and equivalent springs.

As application drawbacks, it can be indicated that this modeling has difficulties in solving structural details such as the indentation between orthogonal walls or the assumption of decoupling of the local failure mode, which requires experience when it comes to irregular arrangements [6].

2.1.2.1 Equivalent beam-based macroelements

The idealization of masonry panels as nonlinear beams represents the most common assumption in the so-called equivalent frame models. Tomaževič [2] proposed a model based on equivalent beams with basic mechanical assumptions where the in-plane damage of masonry facades is due to shear forces in the columns, while the beams and nodal regions are considered rigid and fully resistant. This simple mechanical description, based on simplified elastoplastic relationships, provides sufficient reliability only in the case of weak columns and strong spans. Improvements have been introduced successively implementing the flexibility and limited strength of masonry sills.

Other more advanced equivalent beam-based models have proposed the idealization of the masonry structure as a set of column beam and span beam elements, joined by rigid links representing the nodes between columns and spans (i.e., zones where the seismic damage is rarely observable). These models are based on the phenomenological nonlinear elastoplastic constitutive laws adopted for beam elements.

Another model considers a simple beam for nonlinear analysis of the masonry with two rigid displacements at the ends (simulating the rigid behavior of the intersection of columns and lintel) and a flexible central part. In the Tremuri software, a piecewise linear behavior is incorporated that allows the description of severe damage levels through the progressive degradation of the resistance in correspondence with the floor drift [15].

2.1.2.2 Equivalent spring-based macroelements

Various macroelement models have been formulated by implementing nonlinear springs, within a fictitious frame, to approximate the in-plane nonlinear response of masonry walls and facades (**Figure 4**).

Chen et al. [17] have adapted from reinforced concrete a model with nonlinear shear springs in series with rotational springs for in-plane masonry analysis. This updated model for masonry includes one axial spring, three shear springs, and two rotational springs to simulate failure modes (axial, bed joint slip, diagonal tension, and rocking/crushing) observed experimentally in masonry pillars.

Xu et al. [18] consider the masonry façade as an integral unit in a simple model, using two vertical springs and a nonlinear horizontal spring that governs the shear response of the wall. The hysteretic behavior depends on different parameters, such as the distribution of openings and/or confining elements, relative dimensions, material properties, and boundary conditions of the facade.

2.1.3 Block-based models

Block-based models represent the behavior of masonry at the scale of the main material heterogeneity, characterized by units assembled by mortar (or dry) joints, which governs the main aspects of its mechanical and failure response.

The first example of nonlinear block-based models corresponds to the work of Page [19], where the masonry is considered as an assembly (textured continuum) of elastic brick elements acting in conjunction with connecting elements that simulate the mortar joints that they have limited shear strength depending on the strength of the joint and the level of compression.

This type of model represents the actual union of masonry and structural details, using mechanical parameters obtained from small-scale tests, inclusion of anisotropy, a comprehensible representation of failure modes, and representation of 2D meshes (sheets) and 3D (solids) that allow in-plane and out-of-plane responses to walls and their interactions.

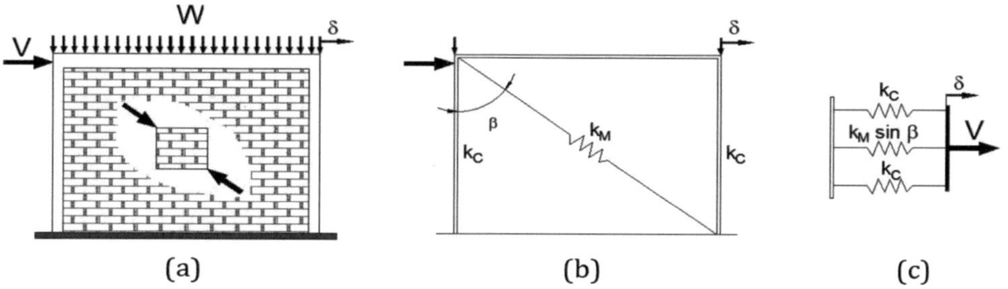

Figure 4.
Equivalent spring-based macroelements [16] (a) Masonry wall subjected to vertical and lateral load, (b) Macromodel, and (c) Simplified model of one degree of freedom with shear strain.

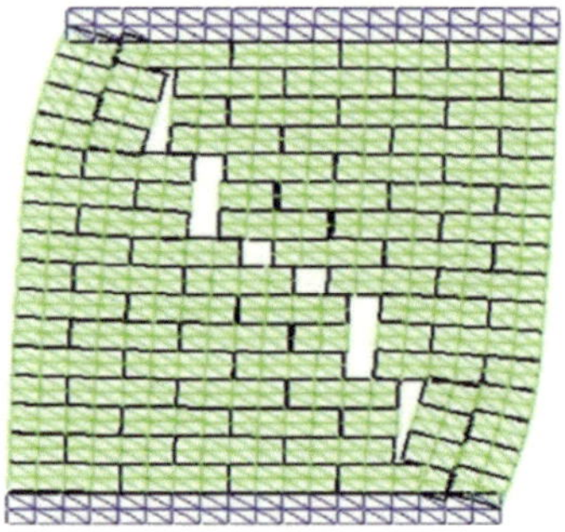

Figure 5.
Scaled deformed mesh obtained from the analysis [20].

The main problem with these models lies in their high computational demand, limiting their applicability at the panel scale. As the actual joint of existing masonry structures is often not fully known, block-by-block discretization could be approximated in those cases. Model assembly is usually a long and complex operation, which limits the use of these modeling strategies to academic studies and a very few high-level consulting groups [6].

Block-based models are classified according to how the interaction between blocks is formulated: on interface elements, on contacts, on textured continuums, on block-based boundary analysis, and on extended finite elements (**Figure 5**) [20].

2.1.3.1 Interface elements

One of the first nonlinear models based on interface elements to simulate the collapse behavior of masonry structures appears in Lofti and Shing [21], where mortar joints are modeled with interface elements of zero thickness and expanded units of masonry (which were considered expanded to take into account the geometry of the mortar joints) were modeled with cracked finite elements. The constitutive model is based on the plasticity of the dilatant interface capable of simulating the initiation and propagation of interface fracture under combined normal and shear stresses.

Lourenço and Rots [22] have developed a multisurface interface-based model in which all nonlinearities (including shear slip, tensile cracking, and also compressive crushing) were concentrated at the interfaces, increasing the efficiency of the model.

The proposal of a cyclic interface model in the mortar joint based on damage mechanics [23] shows a brittle response under tensile stresses and is characterized by frictional dissipation together with stiffness degradation under compressive stresses. In particular, the proposed constitutive equation is based on terms of two internal variables that represent frictional slippage and mortar joint damage. These models are applied for the analysis of 2D problems, which considerably limits the applicability of the modeling strategies to real problems.

2.1.3.2 Contact models

Modeling strategies are based on contact mechanics and are widely used for accurate modeling of masonry structures. Rigid or deformable blocks (linear or nonlinear) interact following a definition of frictional or cohesive-frictional contact. Although several in-house formulations have been developed and validated, three main families of contact-based approaches can be found [6].

1. Discrete Element Methods (DEM) are based on contact penalty formulations and explicit integration schemes implemented in the UDEC (Universal Distinct Element Code). Several applications have been made on actual masonry structures using rigid or linear elastic blocks.

2. An implicit approach that considers the deformability of blocks is Discontinuous Deformation Analysis (DDA). DDA complies with the constraints of no tension between blocks and no penetration of one block into another. Furthermore, Coulomb's law is satisfied in all contact positions for both static and dynamic calculations.

3. Non-Smooth Contact Dynamics (NSCD) method, developed by Jean [24] and characterized by a direct contact formulation, in its non-smooth form, implicit integration schemes, and energy dissipation by block impact. It is applied in dry stone masonry.

None of the approaches can adequately explain the crushing of the masonry, which can be, in some cases, crucial in the mechanical response of these constructions, so other models have been developed that consider the nonlinearity of the block in tension and compression (masonry units and mortar joints as a set of densely packed discrete irregular deformable particles bound together by zero-thickness contact interfaces) [6].

2.1.3.3 Textured continuum models

The main concept of continuous block-based textured models is to model in a context of nonlinear finite elements, masonry, and joints separately without any interface between them, This allows to determine deformations of the two materials, as well as the failure of the blocks, mortar, or mortar joints by adhesion [19].

A continuous block-based textured model discretizes both units and mortar joints with continuous elements, making use of a tension/compression damage model where the damage model has been refined to appropriately reproduce the nonlinear response under shear and to control dilatancy [6].

An innovative approach to mechanically model the nonlinear response of mortar joints from Addessi and Sacco [25], who proposed a micro-structured 3D composite interface formulation based on a multiplane cohesive zone model.

2.1.3.4 Block models based on limit analysis

The limit analysis in the block model allows to accurately and robustly predict the maximum load and its collapse mechanism in masonry buildings. 2D and 3D strategies have been proposed, generally based on limit analysis theorems, even though the effect of friction in calculations is often not an energy-conserving type.

Baggio and Trovalusci [26] proposed a solution of the analysis problem with friction in the interfaces between rigid blocks, that is, they consider the effect of nonlinearity with dilatancy in the solution of the problem.

Ferris and Tin-Loi [27] raised the calculation of the collapse loads of discrete rigid block systems, with unassociated friction and contact interfaces, as a special constrained optimization problem.

On the other hand, Sutcliffe et al. [28] developed a methodology to calculate the loads corresponding to the lower limit in unreinforced masonry walls subjected to shear actions, in plane deformation. Applying the Mohr-Coulomb criterion, the proposed numerical procedure calculates a statically allowable stress field using the finite element method.

Although block-based boundary analysis approaches have also been applied to actual structures such as masonry bridges, their computational demand seems particularly high, which precludes their use for large-scale masonry structures [6].

2.1.3.5 Extended finite element models

Abdullah et al. [29] propose a 3D model that includes a cohesive surface-based behavior to capture the elastic and plastic behavior of masonry joints and a Drucker-Prager plasticity model to simulate masonry crushing under compression.

In addition, XFEM (Extended Finite Element Method) is adopted to model the cracking behavior and compression failure of masonry in infill panels. The discrete interface element is used to simulate the behavior of masonry mortar joints and frame interface joints, showing these approaches as a powerful alternative analysis [30].

2.1.4 Continuum models

The masonry is modeled as a continuous deformable body. The mesh discretization does not have to describe the inhomogeneities of the masonry and can therefore have dimensions that can be larger than the block size. Although there are studies that present an approach at the micromodel level [31] that consist of modeling the masonry units and the mortar as continuous elements, while the masonry-mortar interface is represented by means of discontinuous elements, the scope of this is limited to the study of small specimens.

The computational cost of these continuous macromodel approaches is, in general, less than block-based approaches and much less than micromodels. But the complex behavior of masonry from a mechanical point of view presents a challenge in defining adequate homogeneous constitutive laws.

The parameters to be introduced in the constitutive models can be deduced from experimental tests, or through homogenization techniques, where the constitutive law of the material (considered as homogeneous in the structural scale model) is derived from a homogenization process that relates the scale of the structural model with the scale of a material model (which represents the main heterogeneities of the masonry). The homogenization process is generally based on refined modeling strategies of a representative volume element (RVE) of the structure (**Figure 6**).

2.1.4.1 Direct approach

Direct continuum models are based on continuum constitutive laws that can somewhat approximate the general mechanical response of masonry. Mechanical properties (elastic parameters, strength limits, etc.) can be obtained from experimental tests (**Figure 7**) or other data (for example, analytical or experimentally derived strength domains), without resorting to homogenization procedures based on RVE.

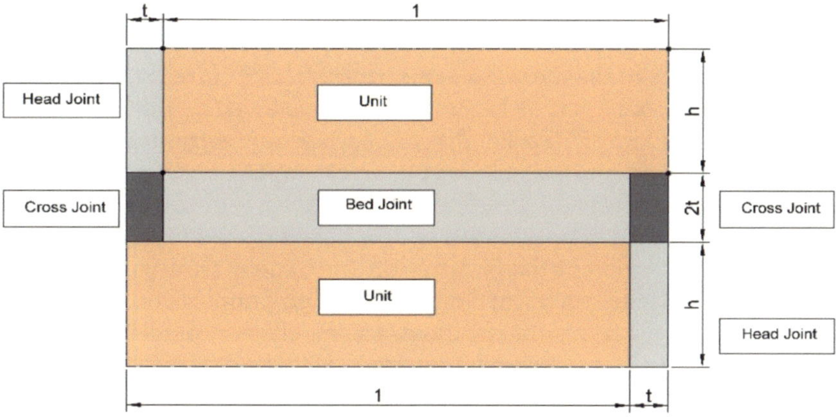

Figure 6.
Basic cell layouts (RVE).

Figure 7.
Preparation of masonry specimens [33].

A first direct approximation consists of an idealization of the mechanical behavior of the masonry. Masonry is conceived as a no-tension material. In general, a material that is not resistant to tensile stress implies an isotropic medium that is unable to withstand these stresses but is also linear elastic. This hypothesis has served as the basis for preliminary structural analyses and has been used in the stability analysis of masonry vaults and domes [13].

Although the cited non-tensile-resisting schemes represent elegant solutions for such a complex problem, their applicability to actual case studies is still limited, to 2D problems, and 3D stress-free structures have only recently been investigated. However, these approaches cannot simulate the post-maximum behavior of masonry structures, which is a strong limitation in the field of seismic evaluation of structures.

In addition, although the zero tensile strength assumption can be considered conservative in general, this could lead to failure mechanisms inconsistent with those observed experimentally, because the masonry tensile strength is not zero.

Other direct continuum models for masonry structures are based on continuous nonlinear constitutive laws based on fracture mechanics (smeared crack models), damage mechanics, or plasticity theory. Various models of smeared cracking [32–34], plasticity, continuous damage and coupled damage, and plasticity have been developed mainly for FEM analysis of concrete structures. However, its usefulness for simulating the collapse or near-collapse behavior of masonry structures has some limitations, mainly due to the multilevel anisotropy (elastic, strength, and brittleness anisotropy) of the masonry and its heterogeneity introduced by the mortar.

The constitutive model of Drucker Prager allows to represent in a simple and easy way the nonlinear behavior of the masonry as an elastoplastic material with dependence on the acting compression, being attractive since it requires the definition of very few parameters that can be determined from diagonal compression tests in the laboratory or the application of flat-jack in situ. To obtain masonry modeling parameters, laboratory tests can be performed on a 1:1 scale on samples of different thickness [35]. With the experimental results obtained, a finite element model is formulated using the ABAQUS software whose parameters allow to be obtained a behavior like that observed during the tests [36].

Although not fully consistent with masonry mechanics, smeared cracking, isotropic damage, and plastic damage models have been widely used to analyze masonry structures, mainly due to their efficiency, their spread in finite element codes, and the relatively few mechanical parameters to characterize the material.

In particular, the use of these models with material nonlinearity has been found to be particularly suitable for the analysis of monumental heritage structures, given their limited computational cost and their ability to represent large-scale and complex 3D geometries. In addition, historic buildings often feature irregular, multilayered masonry, which is not possible to represent block by block and characterize mechanically, moreover, given the strict limitations for in-situ destructive testing of historic buildings of high heritage value [37]. In general, little information is available on the mechanical properties of historic masonry, which favors the use of nonlinear isotropic models.

Many applications with isotropic smeared crack models (isotropic plastic damage) have been carried out successfully in historical towers, churches and temples, palaces, and masonry bridges [6, 38, 39]. Most of the applications in monumental structures are based on 3D models (**Figure 8**), since the structural behavior can rarely be represented by 2D models due the complex and irregular geometries of these buildings.

Although every reliable damage model has to conceive a regularization of the fracture energy, which is normally normalized to a characteristic dimension of the element (characteristic length), very coarse meshes could lead to inaccurate results since their accuracy depends on the strain gradient, the damage pattern and consequently stresses redistribution. An improvement of the constitutive models could be represented using fracture mechanics algorithms, which originate from the analysis of localized fractures in quasi-brittle materials, which ensure mesh independence of numerical results and realistic representation of propagating cracks in the numerical simulation of fracture in quasi-brittle materials [6].

However, when dealing with periodically well-organized masonry, the assumption of a single tensile strength value (governing the tensile response in each direction) runs the risk of being overly simplistic. To this end, some orthotropic nonlinear constitutive laws have been developed and applied in masonry structures. Lourenco et al. have proposed a first example of an orthotropic plasticity model with softening and the ability of that continuous model to represent the inelastic behavior of orthotropic materials to reproduce the resistant behavior of different types of masonry [41].

In recent years, the effect of anisotropy has been introduced through fictitious spaces of isotropic stresses and strains. The properties of the material in the fictitious isotropic space are mapped to the real anisotropic space by means of a consistent

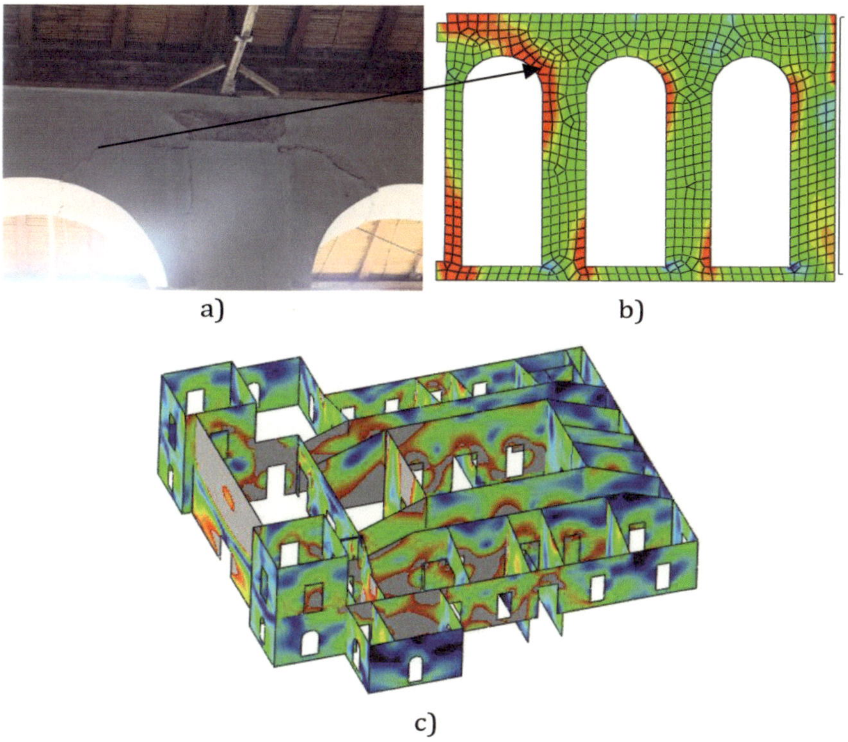

Figure 8.
Finite element model of the damaged structure [35, 40]: (a) Damage status, (b) stress state of the damaged sector, and (c) stress state of the masonry due to non-homogeneous settlements.

fourth-order operator. It has the advantage that classical plasticity theory can be used to model nonlinear behavior in anisotropic spaces [42, 43].

From this concept, an orthotropic damage model has been developed specifically for the analysis of masonry subjected to in-plane cyclic loading. Different elastic and inelastic properties are adopted along the two natural axes of the masonry (i.e., the directions of bed joints and head joints) also as principal axes of damage, since when stresses are reversed, the crack closes, and the material regains its stiffness.

Martín proposes an anisotropic damage model that decouples the behavior in tension and compression, in addition to contemplating the directionality of the damage [44].

Pelà et al. [45] have more recently proposed an orthotropic damage model for masonry analysis, in which the orthotropic behavior is simulated through mapping tensors that link the real anisotropic field with an auxiliary fictitious space. The model allows the simulation of orthotropic-induced damage, while accounting for unilateral effects, through a decomposition of the stress tensor into tensile and compressive contributions. The damage model has also been combined with a crack tracking technique to reproduce localized crack propagation in the FE problem [6].

Although direct continuum anisotropic approaches represent scientifically sound solutions, their application in real cases is scarce due to their computational cost and fundamentally to the number of properties of the material to be mechanically characterized, which is substantially higher than isotropic approaches.

2.1.4.2 Homogenization procedures and multi-scale approaches

A constitutive law of a homogeneous model at the structural scale that tries to represent the masonry can be deduced from the homogenization processes based on RVE. The definition of an adequate RVE is crucial, since it must be representative of the heterogeneity of the scale of the material under study, incorporating the characteristic heterogeneities of the material in a statistical way. Various RVE geometries have been proposed, to account for different periodic and non-periodic masonry patterns (**Figure 6**).

Given the mechanical complexity of masonry, in terms of anisotropy, three main families of approaches can be distinguished [6]:

- a priori homogenization where first an RVE-based homogenization is performed to deduce the properties of the material at the structural scale and then the homogenized mechanical properties are introduced into the model at the structural scale,

- step-by-step multi-scale where the general behavior at the structural scale is determined step by step by solving a boundary value problem (BVP) in the RVE for each integration point of the model at the structural scale and from that determination, an average response is estimated as a constitutive relationship in the step-by-step structural scale model.

- adaptive multi-scale, in which the material scale model is adaptively inserted and resolved into the structural scale model, thus establishing a strong coupling between the two scales.

2.2 Confined masonry modeling

In this case, there is a combination of elements of different materials with different physical-chemical-mechanical properties. In general, vertical and horizontal ties of reinforced concrete or steel are used, forming a framework resistant to vertical and horizontal loads, closing its openings with masonry.

The experiences obtained from earthquakes and from laboratory tests have shown a different behavior when the resistant framework has masonry than when it is not filled.

The interaction mechanisms between the infill masonry and the reinforced concrete frame system may require two design approaches [46]:

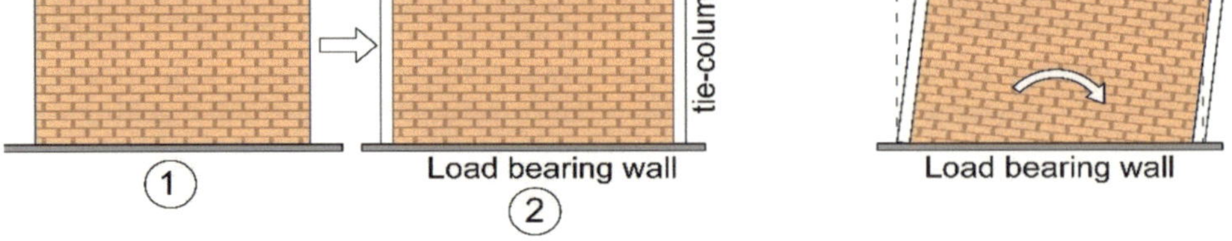

Figure 9.
Constructive process of confined masonry and behavior under seismic loads.

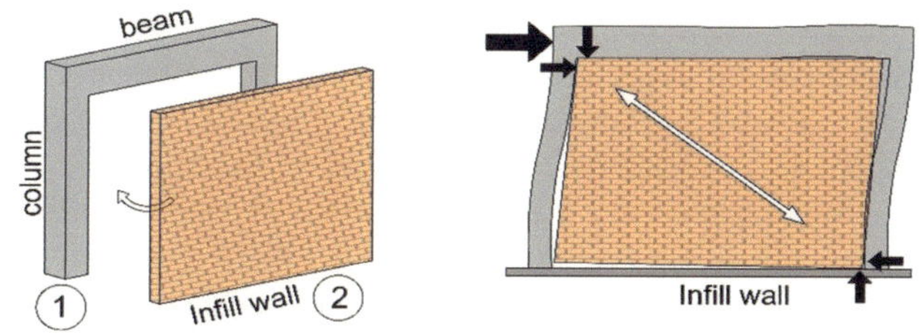

Figure 10.
Constructive process of filled confined masonry and behavior under seismic loads.

- The wall infill is constructed as a constituent part of the structural system. In this case, the effect of interaction of forces between the masonry filling and the reinforced concrete framework must be considered (**Figure 9**).

- The wall fill is built as a secondary structural element, separated from the main structure by means of suitable joints, allowing the main structure to deform freely during the earthquake (**Figure 10**).

2.2.1 Confined masonry with box behavior

Brick masonry buildings have a great mass and, therefore, large horizontal forces are generated during an earthquake, causing damage due to shear, tensile, and compressive stresses. A proper choice of structural configuration helps to minimize damage and prevent collapse. Earthquake evidence shows that confined masonry constructions with adequate wall density can withstand major earthquakes without collapse [46].

The most appropriate structural model is identified as "box action" that connects reinforced concrete beams and columns with masonry panels, floors, and ceilings. The horizontal beams at the plinth, parapet, lintel, and gable level support the masonry walls forming a unit. At ceiling level, flat or inclined reinforced concrete, ceramic, or wood slabs can also be used. Poorly connected roof or unduly thin walls are threats to good seismic performance.

The earthquake-resistant construction regulations have been incorporating the confined masonry design for use in social housing or buildings with symmetrical floors, considering the general shape and size of the building with limitations of slenderness and heights, the distribution of weight and elements resistant to lateral load in a regular and symmetrical way throughout the considered building [3, 14, 47].

2.2.2 Confined masonry with equivalent structures

Infill masonry significantly increases the rigidity of the structural system; therefore, to determine the interaction forces between the framework and its infill, it is necessary to know the contribution of the constituent elements to the lateral resistance of the assembly, as well as the change in the contribution with the increase in inelastic deformations of the assembly during the earthquake.

The investigations carried out by Zarnic and Tomaževič in the 1980s [2] have made it possible to evaluate this behavior, and their results have been incorporated into different regulations on earthquake-resistant constructions, especially for their application in the construction of social housing [47].

To model the inclusion of masonry in reinforced concrete frames, a fictitious compression diagonal can be used. Due to the complexity of the behavior of the structural system, the simplified numerical model must be based on the results of quasi-static and cyclic dynamic tests [3]. The layout of the compression diagonal is affected by location, size, and slope in such a way that it must be adjusted to achieve the combined structural behavior of the fill and the limiting structure.

The lateral stiffness in the plane of the filled frame is different from the sum of the independently interacting elements. Tests have shown that under seismic loads, the reinforced concrete structure separates from the fill, reducing the initial lateral stiffness due to nonlinear behavior of the system and reaching 60% of the maximum seismic load [3, 48].

The American standard ASCE/SEI 41-13 guides on how to model the diagonal compression strut in the structural system with different arrangements: concentric (**Figure 11a**), eccentric (**Figure 11b**), at an angle of 45° (49), or in combination when they present openings (**Figure 11c**) [3]. The criteria for dimensioning the diagonals for the calculation model maintain the same thickness as the masonry panel as thickness and its height is a function of the width of the panel [47–49].

For undamaged infill panels, the arching effect of the masonry provides significant resistance to out-of-plane forces. This effect decreases when the filler is damaged due to in-plane forces. The exact mechanisms of deterioration cannot be reliably quantified, and therefore, the two actions are currently considered separately [5].

2.3 Modeling of reinforced masonry

Reinforced masonry with distributed reinforcement is one in which there is horizontal and vertical reinforcement distributed throughout the wall, placed in such a way that the masonry, mortar, concrete, and steel act together to resist the stresses. In this type of masonry, the placement of confined columns is not necessary (**Figure 12**) [46].

The presence of vertical and horizontal reinforcement in ceramic or concrete masonry units improves the resistance and ductility of the resistant wall. The vertical reinforcement is positioned in the hollow cores of the masonry unit where concrete is injected to anchor the reinforcement and protect it from corrosion according to the calculation of the reinforcement section necessary to absorb the stresses. The

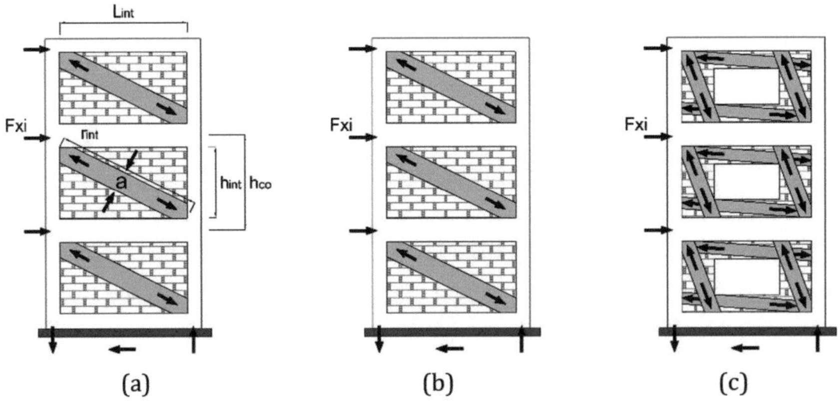

Figure 11.
(a) Concentrically located compression strut analogy, (b) eccentrically located compression strut analogy, (c) compression strut analogy in infill walls with openings.

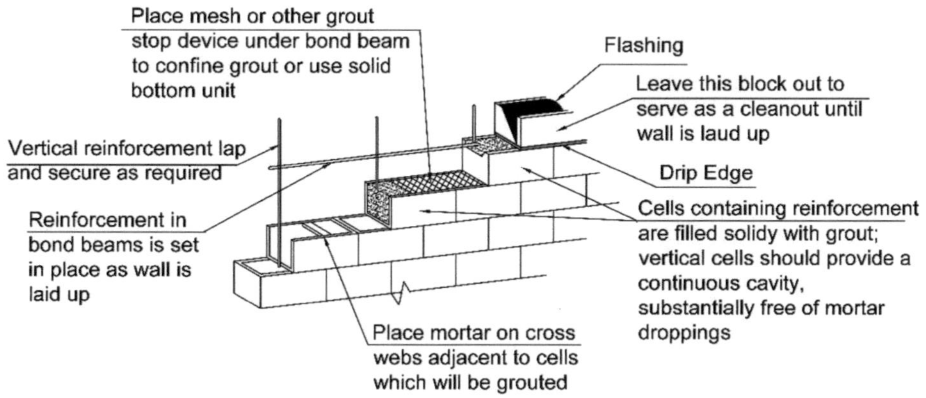

Figure 12.
Placing reinforcement in hollow bricks in a masonry wall [9].

horizontal reinforcement is located in the horizontal joints or in the connection beams of the floor and lintel. The amount of reinforcement is calculated as reinforced concrete based on the acting loads [3, 47].

3. Conclusions

The service life of well-constructed masonry structures over time indicates that it is a sustainable material due to its durability, adaptability, and maintainability. However, current construction practices of production with great speed and minimum amount of material impact the traditional image of this solid, durable, and sustainable construction material.

Finite element modeling has had a great evolution in applications to masonry structures, with different degrees of difficulty depending on the type of masonry, layout or rigging, edge conditions, data availability and tests, taking into account the objective of its application and experience level of the modeler.

Finite element method modeling for historical masonry structures is considered to have made great progress in the last decade, and the different software available adapts to the different conformations of the masonry structure. Improvements are still pending regarding connectors, sealants in joints and behavior of coatings under different environmental conditions.

Finite element models and structural element models represent the behavior of the masonry at different scales, therefore the predictions of the behavior of the masonry present differences, which in some cases can be significant.

Confined masonry modeling is based on field and laboratory experiences. The design guidelines present the current earthquake resistant regulations, either considering the box behavior or as an analogy of compression struts.

The modeling of reinforced masonry applies criteria of reinforced concrete macro-models.

Practical models of masonry structures available in masonry structure codes apply to current construction guidelines.

Author details

Gerardo González del Solar, María Domizio, Pablo Martín and Noemi Maldonado*
National Technological University, Mendoza, Argentina

*Address all correspondence to: ngm@frm.utn.edu.ar

References

[1] Ciblac T, Morel J-C. Sustainable masonry. In: Stability and Behavior of Structures. 1st ed. London: ISTE LTD and Wiley; 2014. p. 287. DOI: 978-1-84821-495-8

[2] Tomaževič M. Earthquake-Resistant Design of Masonry Buildings. London: Imperial College Press; 2006. p. 282. DOI: 1-86094-066-8

[3] ASCE standard ASCE/SEI 41-13. Chapter 11: Masonry. In: American Society of Civil Engineers: Seismic Evaluation and Retrofit of Existing Buildings. 2013. pp. 225-250. DOI: 978-0-7844-7791-5

[4] Maldonado N, Martín P, González del Solar G, Domizio M, Maldonado I. Puesta en Valor de la Mampostería Histórica: Casos de Estudio en Mendoza, Argentina. Revista Tecnología y Ciencia. 2020;**39**:103-117. DOI: 10.33414/rtyc.39.103-117.2020

[5] Maldonado N, Martín P, Maldonado I, Domizio M, González del Solar G, Calderón F. Behaviour and durability of ceramic heritage masonry in near source fault zone. In: Proceedings of 16th World Conference on Earthquake (16WCEE). Santiago de Chile. Chile: IAEE; 2017. p. 3371

[6] D'Altri A, Sarhosis V, Milani G, Rots J, Cattari S, Lagomarsino S, et al. Modeling strategies for the computational analysis of unreinforced masonry structures: Review and classification. Archives of Computational Methods in Engineering. 2020;**27**(4):1153-1185. DOI: 10.1007/s11831-019-09351-x

[7] Lagomarsino S, Cattari S. PERPETUATE guidelines for seismic performance-based assessment of cultural heritage masonry structures. Bulletin Earthquake Engineering. 2015;**13**(1):13-47. DOI: 10.1007/s10518-014-9674-1

[8] Válek J, Hughes J, Groot C. Historic Mortars. Characterisation, Assessment and Repair. RILEM Book Series 7. Dordrecht, Heilderberg, New York, London: Springer; 2012. p. 444. DOI: 10.1007/978-94-007-4635-0

[9] Klingner R. Masonry Structural Design. New York: McGraw-Hill; 2010. p. 589. DOI: 978-0-07-163831-9

[10] Lourenço P. Computational Strategies for Masonry Structures [Thesis]. The Netherlands: Delft University of Technology; 1996

[11] Asteris P, Cotsovos D, Chrysostomou C, Mohebkhah A, Al-Chaar G. Mathematical micromodeling of infilled frames: State of the art. Engineering Structures. 2013;**56**:1905-1921. DOI: 10.1016/j.engstruct.2013.08.010

[12] Alforno M, Venuti F, Monaco A, et al. Seismic behaviour of cross vaults with different brick pattern. Bulletin of Earthquake Engineering. 2022;**20**:3921-3939. DOI: 10.1007/s10518-022-01347-6

[13] Heyman J. The stone skeleton. International Journal of Solids and Structures. 1966;**2**(2):270-279. DOI: 10.1016/0020-7683(66)90018-7

[14] Ordinanza del Presidente del Consiglio dei Ministri (OPCM). Norme tecniche per il progetto, la valutazione e l'adeguamento sismico degli edifici; 2005

[15] Lagomarsino S, Penna A, Galasco A, Cattari S. Tremuri program: An equivalent frame model for the

nonlinear seismic analysis of masonry buildings. Engineering Structures. 2013;**56**:1787-1799. DOI: 10.1016/j.engstruct.2013.08.002

[16] Sánchez S, Arroyo R, Jerez S. Modelo de un grado de libertad para evaluar la curva carga lateral-distorsión en muros de mampostería confinada. Revista de Ingeniería Sísmica. 2010;**83**:25-42. DOI: 10.18867/ris.83.143

[17] Chen S, Moon F, Yib T. A macroelement for the nonlinear analysis of in-plane unreinforced masonry piers. Engineering Structures. 2008;**30**:2242-2252. DOI: 10.1016/j.engstruct.2007.12.001

[18] Xu H, Gentilini C, Yu Z, Wu H, Zhao S. A unified model for the seismic analysis of brick masonry structures. Construction and Building Materials. 2018;**184**:733-751. DOI: 10.1016/j.conbuildmat.2018.06.208

[19] Page A. Finite element model for masonry. Journal of the Structural Division. 1978;**104**(8):1267-1285. DOI: 10.1061/JSDEAG.0004969

[20] Chaimoon K, Attard M. Modeling of unreinforced masonry walls under shear and compression. Engineering Structures. 2007;**29**:2056-2068. DOI: 10.1016/j.engstruct.2006.10.019

[21] Lotfi H, Shing P. Interface model applied to fracture of masonry structures. Journal of Structural Engineering. 1994;**120**(1):63-80. DOI: 10.1061/(ASCE)0733-9445(1994)120:1(63)

[22] Lourenço P, Rots J. Multisurface interface model for analysis of masonry structures. Journal of Engineering Mechanics. 1997;**123**(7):660-668. DOI: 10.1061/(ASCE)0733-9399

[23] Gambarotta L, Lagomarsino S. Damage models for the seismic response of brick masonry shear walls. Part I: The mortar joint mode and its applications. Earthquake Engineering and Structural Dynamics. 1997;**26**(4):423-439. DOI: 10.1002/(SICI)1096-9845(199704)26:4%3C423::AID-EQE650%3E3.0.CO;2-#

[24] Jean M. The non-smooth contact dynamics method. Computer Methods Applied Mechanics Engineering. 1999;**177**(3-4):235-257. DOI: 10.1016/S0045-7825(98)00383-1ff. ffhal01390459f

[25] Addessi D, Sacco E. Nonlinear analysis of masonry panels using a kinematic enriched plane state formulation. International Journal Solids Structure. 2016;**90**:194-214. DOI: 10.1016/j.ijsolstr.2016.03.002

[26] Baggio C, Trovalusci P. Limit analysis for no-tension and frictional three-dimensional discrete systems. Journal Structural Mechanics. 1998;**26**(3):287-304. DOI: 10.1080/08905459708945496

[27] Ferris M, Tin-Loi F. Limit analysis of frictional block assemblies as a mathematical program with complementarity constraints. International Journal of Mechanical Sciences. 2001;**43**(1):209-224. DOI: 10.1016/S0020-7403(99)00111-3

[28] Sutcliffe D, Yu H, Page A. Lower bound limit analysis of unreinforced masonry shear walls. Computers & Structures. 2001;**79**(14):1295-1312. DOI: 10.1016/S0045-7949(01)00024-4

[29] Abdulla K, Cunningham L, Gillie M. Simulating masonry wall behaviour using a simplified micro-model approach. Engineering Structures. 2017;**151**:349-365. DOI: 10.1016/j.engstruct.2017.08.021

[30] Zhai C, Wang X, Kong J, Li S, Xie L. Numerical masonry-infilled rc frames using xfem. Journal of Structural Engineering. 2017;**143**(10):04017144. DOI: 10.1061/(ASCE) ST.1943-541X.0001886

[31] Sánchez A. Refuerzo de Muros de mampostería de gran espesor para zona sísmica de alta sismicidad: análisis no lineal mediante aplicación de superficies de interacción [thesis]. Mendoza, Argentina: Universidad Tecnológica Nacional Facultad Regional Mendoza; 2014

[32] González del Solar G, Martín P, Maldonado N. Formulation, implementation and validation of a scalar damage model for brittle materials applied to three-dimensional solid elements. Revista Ingeniería de Construcción RIC. 2018;**33**(1):111-122. Available from: http://www.ricuc.cl

[33] González del Solar G. Modelo de Daño Escalar para Muros de Mampostería de Ladrillo Macizo Cocido con una Junta Vertical en el espesor [thesis]. Mendoza, Argentina: Universidad Nacional de Cuyo; 2022

[34] Hillerborg A, Modéer M, Petersson P. Analysis of crack formation and crack growth in concrete by means of fracture mechanics and finite elements. Cement & Concrete Research. 1976;**6**(6):773-781. DOI: 10.1016/0008-8846(76)90007-7

[35] González del Solar G, Martín P, Calderón F, Maldonado N, Maldonado I. Importancia de la modelación numérica en la puesta en valor de estructuras patrimoniales de mampostería en zona sísmica. Revista ALCONPAT. 2014;**4**(3):215-231. DOI: 10.21041/ra.v4i3.71

[36] Hibbitt H, Karlsson B, Sorensen P. Abaqus Theory Manual. Vol. V.5.8. USA: Dassault Systèmes Simulia Corp; 2011

[37] D'Altri AM, Castellazzi G, de Miranda S. Collapse investigation of the Arquata del Tronto medieval fortress after the 2016 Central Italy seismic sequence. Journal of Building Engineering. 2018;**18**:245-251. DOI: 10.1016/j.jobe.2018.03.021

[38] González del Solar G, Orrego J. Estudio del comportamiento de muros de mampostería simple de gran espesor solicitados biaxialmente [thesis]. Mendoza, Argentina: Universidad Tecnológica Nacional Facultad Regional Mendoza; 2013

[39] Maldonado N, Martín P, Maldonado I. Seismic mitigation of a historic masonry building. The Open Construction and Building Technology Journal. 2011;**5**(I-M3):61-70. DOI: 10.2174/1874836801105010061

[40] Maldonado N, Martín P, Maldonado I, Calderón F, González del Solar G, Domizio M. Estudios para la puesta en valor de un edificio patrimonial con pinturas murales en zona sísmica: un caso de estudio. Revista Internacional Tech ITT. 2016;**38**(14):4-15

[41] Lourenço P, De Borst R, Rots J. A plane stress softening plasticity model for orthotropic materials. International Journal of Numerical Methods Engineering. 1997;**40**(21):4033-4057. DOI: 10.1002/(SICI)1097-0207(19971115)40:21%3C4033::AID-NME248%3E3.0.CO;2-0

[42] Luccioni B. Formulación de un Modelo Constitutivo para Materiales Ortótropos [thesis]. Tucumán, Argentina: Universidad Nacional de Tucumán; 1993

[43] Luccioni B, Oller S. A directional damage model. Computer Methods in Applied Mechanics and Engineering.

2003;**192**(9-10):1119-1145. DOI: 10.1016/S0045-7825(02)00577-7

[44] Martín P. Modelo de daño anisótropo [thesis]. Tucumán, Argentina: Universidad Nacional de Tucumán; 2001

[45] Pelà L, Cervera M, Roca P. An orthotropic damage model for the analysis of masonry structures. Construction and Building Materials. 2013;**41**:957-967. DOI: 10.1016/j.conbuildmat.2012.07.014

[46] Earthquake Engineering Research Institute. In: Meli R, Brzev S, editors. Seismic Design Guide for Low-Rise Confined Masonry Buildings. Oakland, California: EERI; 2011. p. 91. ISBN: 978-1-932884-56-2

[47] INPRES-CIRSOC Editors. Reglamento Argentino para Construcciones Sismorresistentes. Parte III Construcciones de Mampostería. 1st ed. INTI: Buenos Aires; 2018. p. 73

[48] Michelini R, Maldonado N, Pizarro N, Olivencia L. Análisis experimental de la interacción de tabiques de hormigón armado con mampostería para diseño estructural sismorresistente. XXIX Jornadas Sudamericanas de Ingeniería Estructural. Punta del Este. Uruguay. 2000;**1**:60

[49] Stavridis A, Shing P. Finite-element modeling of nonlinear behavior of infilled RC frames. Journal of Structural Engineering. 2010;**136**(3):285-296

Section 2

The Physical Properties of Masonry

Chapter 4

Compressive Strength Test of Interlocked Blocks Made with High-Mechanical-Performance Mortars

Edrey Nassier Salgado Cruz, Alberto Muciño Vélez, Eligio Alberto Orozco Mendoza and César Armando Guillén Guillén

Abstract

Conventional masonry pieces are simple construction elements used for the building of houses for a long time. Nevertheless, the rapid growth in the demand for social and middle-class housing in developing countries has forced engineers to develop cheaper and new creation processes and systems with better features and qualities. In this sense, to obtain an optimization in masonry pieces, the following must be considered: 1) the material from which it is fabricated and 2) the design (shapes and geometry). As an alternative, in this work, we present the design of interlocked concrete blocks with measures of 12.5 cm wide, 25 cm in height, and 40 cm in length, made with mortar mixtures with high mechanical performance, with which wall sections were built (masonry assemblies of 62.5 × 60 × 12.5 cm and prisms of 62.5 × 40 × 12.5 cm) and then characterized according to standards of the mechanical compression tests. The obtained mechanical compressive strengths were 177.72 kg/cm^2 in the unitary masonry pieces, 47.4 kg/cm^2 in prisms, and 3.98 kg/cm^2 in diagonal compression tests. This type of masonry materials and their assembly procedure can be useful for the manufacture of middle-income and social housing in developing countries.

Keywords: mortarless, high-performance mortars, masonry, interlocked blocks, strength test

1. Introduction

Nowadays, masonry units represent a large part of the constructed surroundings, it is estimated that 80% of the world's population lives in houses built with this type of construction material [1]. The fabrication of structural walls, made by assembling masonry pieces such as clay bricks and concrete blocks using a mortar bed between 10 and 20 mm, it is normally the most used system for the construction of low-income and social (middle-income) housing in the Metropolitan Area of Mexico.

The popularity of masonry constructions is due, among other aspects, to its low cost, the local availability of the necessary materials, and the use of traditional construction techniques. Masonry blocks can be made from a variety of materials, types, dimensions, and can be placed in different ways, normally using a thin mortar layer ($\approx$ 2 cm) that allows linking the blocks. The union mortar can also be made with different types of conglomerates and sand, mixed with water. The mechanical properties of blocks and mortars are very different due to the components that constitute them, from its manufacturing process, its geometry, and size; therefore, when a set of masonry blocks united by a series of mortar layers is subjected to a compression load, a complex interaction appears in the transition zones.

Thus, lately, it has been decided for the development of new systems, methods, and procedures to build masonry, to try to eliminate some of the disadvantages of traditional methods in wall construction, for this, interlaced masonry blocks with dry piled up system have been explored; it is well-known as mortarless masonry technology. This technology replaces conventional masonry construction by eliminating mortar layers, for masonry units with special characteristics in their geometry, which are mechanically interconnected through slots, unions, or tongue pieces that facilitate interlocking and load transfer [2].

Interlocked blocks have the advantage of accelerating masonry construction and/or improving the structural behavior of walls [3]. The interlaced masonry systems reported [4] are varied and adapted in terms of their mechanical performance, and most of these systems have being patented [5]. Currently, there are no specific construction regulations applicable to these systems, so they are governed by the constitutive laws of traditional masonry. Some masonry systems with interlocking blocks reported in the literature are the following: Mecano system [6], Sparfil [7], Haener [8], modified Hblock [9], Sparlock [10], WHD block [11], the Solbric and Hydraform systems [12], the Bamba, Auram, and Tanzanian systems [13], among others.

These systems have a series of advantages with respect to traditional masonry processes, such as the positioning of the blocks being simple and requiring less skilled labor. With this, the construction of the walls is faster, which leads to higher productivity [14, 15]. As they are done without a mortar bed, construction inaccuracies due to manual work are eliminated and problems associated with the specific properties of the mortar used to join the masonry blocks are also overcome [13]. In addition to facilitate the construction processes [16, 17], the technical use of these aids to promote sustainable construction [18] and the construction of structures with high mechanical resistance [19–21] that have been successfully investigated in areas of high seismicity [22–25].

The constructive system based on the use of interlocking masonry blocks has been widely used in several developing countries in such a way that it has gradually replaced conventional masonry procedures [26]. On the other hand, in Mexico, many of the masonry pieces that are currently used in conventional constructive processes do not meet the minimum requirements related to mechanical performance established by the local norms, due to the lack of control in the manufacture of the pieces and the erroneous use of materials for its manufacture [27–31]. Furthermore, interlocked block/based construction methods in Mexico are very few. In this sense, the intention of this investigation, in the first term, is to show the process to develop a high-mechanical-performance mortar and its implementation in the manufacture of interlocking masonry pieces, capable of being used in construction systems, and the second, the design, construction, and testing of the mechanical resistance of the manufactured blocks.

2. Design and testing of the mortar used to manufacture the interlocked blocks

2.1 Materials used for the manufacture mortars

The cement used was Portland 30 R type, which is widely commercialized in the center of the Mexican Republic, mainly in the State of Mexico, Mexico City, Puebla, Hidalgo, and Querétaro, which makes it one of the most used brands in the construction and auto-construction. The fine aggregates for the preparation of the mixtures were sands of "pink color" obtained from the quarries of the volcanic zone to the east of Mexico City, whose mineralogical composition is described in the reference (**Table 1**) [32].

To make the mixtures, drinking water from the municipal network was used, a plasticizer additive, and finally, polypropylene microfiber, another additive, whose main function is to act as a secondary reinforcement of the mortar.

2.2 Sands characterization

A granulometric analysis was carried out on the sands, which is established in the ASTM- C 136 standard [30]. The process consisted of mechanically separating an aggregate sample (200 grams), previously dried in the oven at 110 $\pm$ 5°C [31], through a series of sieves with openings established progressively smaller than the norm, with the intention of determining the sizes and the gross weight of each size with respect to the total number of particles.

The data obtained from this analysis are represented in the form of a curve, where the percentage of weight that passes through the mesh is plotted on the ordered axis and the diameter of particles on the abscissa. **Figure 1** shows the granulometric curve of the sand used in the experimentation.

2.3 Mortar mixtures

According to what is established for high-mechanical-performance concrete by the American National Institute of Science and Technology (NIST) and the American Concrete Institute (ACI), high-performance concrete is homogeneous, made with

Name of the Phase/Data base of reference	Chemical formulas
Albita [30]	$Na_{0.983}Ca_{0.012}K_{0.005}AlSi_3O_8$
Low Albita [31]	$NaAlSi_3O_8$
Anortita sodiana [32]	$Na_{0.48}Ca_{0.52}Al_{1.52}Si_{2.48}O_8$
Celedonita [33]	$K(Mg_{0.78}Fe_{1.22})Si_4O_{10}(OH)_2$
Cuarzo [34]	SiO_2
Richterita [35]	$Na_{1.62}Mg_{6.19}Si_3(0_{22})(OH)_2$
Sanidina [36]	$KFe_{0.5}Si_{3.0}1Al_{0.49}O_8$

As reported by Muciño A. et. to. [32], the "pink" sand used in the elaborated mortar mixtures has the following mineralogical composition [32].

Table 1.
Mineral identification of the aggregate – pink sand.

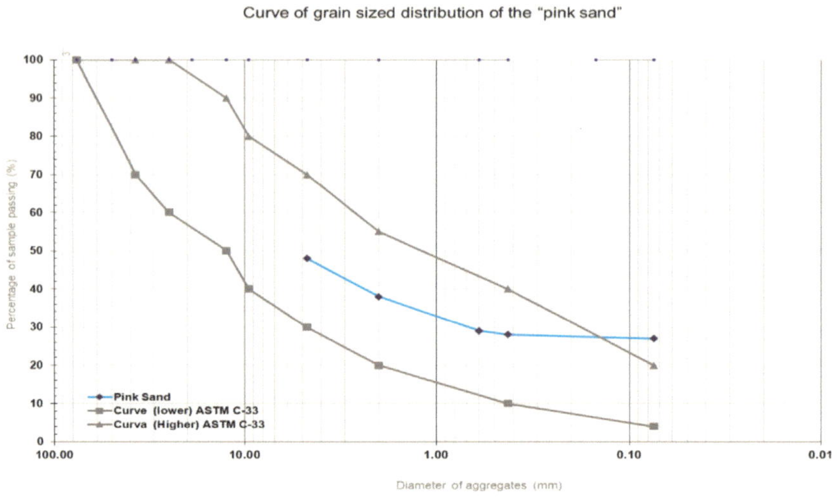

Figure 1.
Grain-sized distribution of the "pink" sand used in the manufacture of mortars.

high-quality components, with good adherence, without segregation, and its mechanical properties must be stable and with high early strength [33].

Therefore, cement mixtures were made with water, sand, and additives following the ASTM C-109 standard trying to obtain the physical and mechanical properties of a high-performance mortar, by implementing best practices and mixing designs, based on the aforementioned definition. The ASTM C109 standard describes the test method and the necessary conditions for the determination of the compressive strength of hydraulic cement mortars in cubic specimens with a side of 2 inches [34, 35].

Table 2 shows the quantities of the materials used for the preparation of five cubic specimens (5×5×5 cm) for each type of mixture (M1, M2, and M3), used for each day of the compressive strength test (1, 3, 7, 14, and 28 days). All the mixtures were prepared in a ratio of 1:2.75 (cement-sand), as follows: the first mixture, M1, was taken as a reference and was prepared with cement-sand and drinking water from the municipal network, while M2 mixture was added with a superplasticizer as an additive, and M3 mixture, apart from the additive, was also added with polypropylene microfiber.

For each mixture and by every day of test five buckets were used, with a volume by 125 bucket of cm3, for 3125 a total volume of cm3 by the 25 buckets, for each type of mixture and by every day of test (column 2), for this volume 1641.5 grams of

Type of mortar	Volume	Cement	Sand	Water	Proportion	-	Plasticizer	Polypropylene microfibers
	cm3	Grams	Grams	Grams	ASTM C-109	w/c	Grams	Grams
M1	3125.0	1641.5	4514.1	1477.4	1: 2.75	0.90	—	—
M2	3125.0	1641.5	4514.1	574.4	1: 2.75	0.35	528.8	—
M3	3125.0	1641.5	4514.1	574.4	1: 2.75	0.35	528.8	31.9

Table 2.
Mortar mixtures with additives (M2–M3) and without additive (M1).

cement (column 3) and 4514.1 grams of sand were used (column 4), the M1 mixture required 1477.4 grams of water so that the consistency was fluid and workable, for mixture 3 and 4, 574.4 grams of water was used to obtain the same workability and fluidity that the M1 mixture (column 5), obtaining a relation cementitious water of 0.9, 0.35, and 0.35 for mixes M1, M2, and M3, respectively (column 7), to the mixture M2 and M3 528,8 grams of superplasticization additive (column 8) was added, and finally to the M3 mixture, aside from the plasticizer, 31.9 grams of microfibers of polypropylene were added (column 9).

The amount of the materials used was established employing a previous sampling of the mixtures to obtain the optimal quantity of additives in each one of them. Thus, in the fresh state, the parameters considered as essential were fluidity and setting, while in the hard state, it was the compressive strength.

2.4 Compressive strength of mortar

The manufacturing process was as follows: the mortars materials were mixed manually and spilt into molds of established dimensions (50 mm by side). Subsequently, they were subjected to mechanical vibrations to guarantee the release of possible air bubbles. After 24 hours, the samples were removed from molds and left to stand at room temperature. To ensure the perpendicularity between faces, the samples were treated under a mechanical rectified process. So, the treatment guaranteed a homogeneous distribution of stresses during the compression tests. The compressive strength tests were carried out in an INSTRON hydraulic compression machine (model 400RD-E1-H2) at a load rate of 1 kN/second until rupture [34].

The compression at break was obtained from the average of 3–5 tested cubes, declaring the relationship between the total load supported during the test and the contact area of the cube section. **Figure 2** shows the compressive strength values at 28 days for the three mixtures.

In this research, for the manufacture of the blocks, the M3 mixture (cement, sand, water, plasticizer, and micro polypropylene fibers) was selected, due to the ductile behavior of the material, when reaching the breaking point, the mortar bucket was not disintegrated, as the M1 and M2 mixtures (**Figure 3a** and **b**), and the time to reach its last state of failure was greater, being an important aspect in the performance of a structural element.

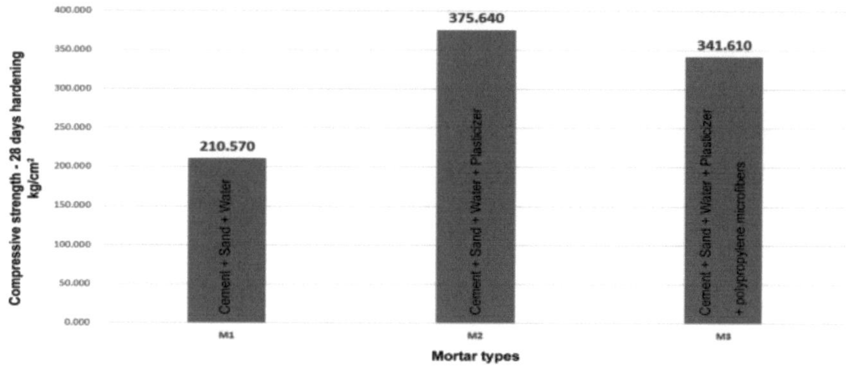

Figure 2.
Maximum breaking strength in compression test for mixtures M1, M2, and M3 at 28 days of hardening.

Figure 3.
Samples after the compressive strength test. Mixtures: a) M1; reference mortar, b) M2; mortar with plasticizer, c) M3; mortar with microfibers and plasticizer.

3. Interlocking block system design

The block designed with high mechanical resistance to compression mortar is a solid piece since it has a net area >75% of the gross area and its internal and external walls have a thickness of 30 mm (**Figure 4a**), fulfilling the requirements established in NMX-C-404-ONNNCCE norm [36]. The block is prismatic in shape and has a smooth face, is made up of two pieces, a female piece, and a male piece (**Figure 4b**) that allows mechanical union, in such a way that the use of impact mortar to join them is avoided.

3.1 Manufacture of interlocking blocks

The elements of the interlocking blocks (**Figure 5**) made by manual draining of the M3 mixture described in **Table 2** were made in wooden molds with the appropriate dimensions (**Figure 6**). After 10 days of air drying under ambient conditions

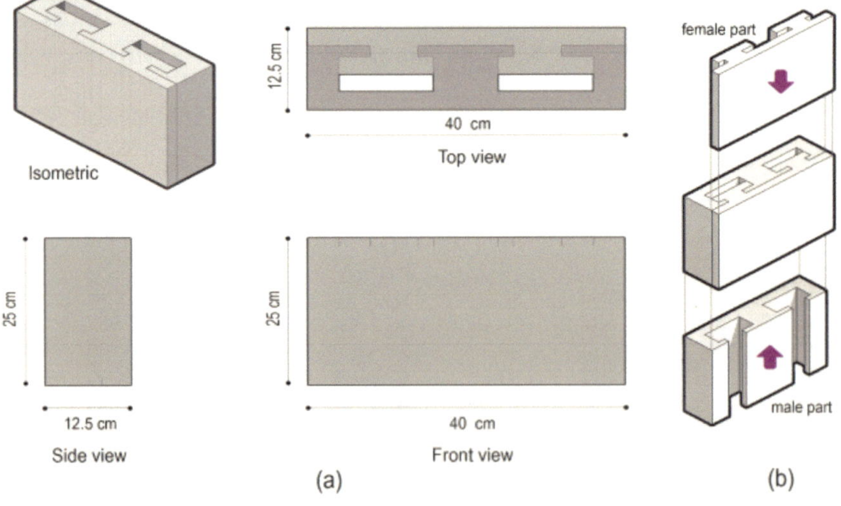

Figure 4.
a) Isometric and plans of the designed block, with dimensions of 40 centimeters long, 25 centimeters high, and 12,5 centimeters wide. b) Form of union of the block that consists of two pieces, female piece, and male piece.

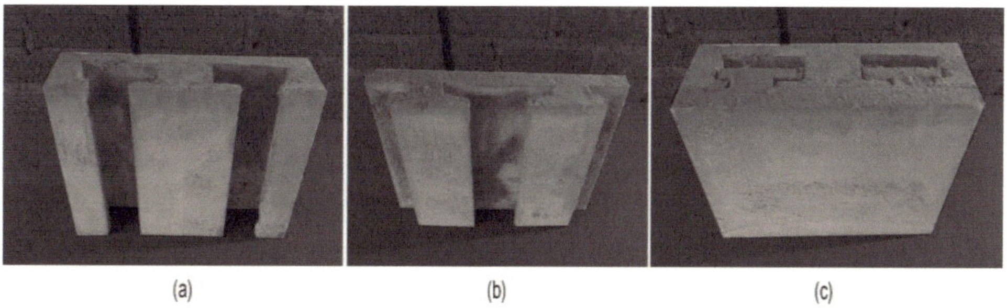

Figure 5.
a) Male piece, b) female piece, c) assembled block.

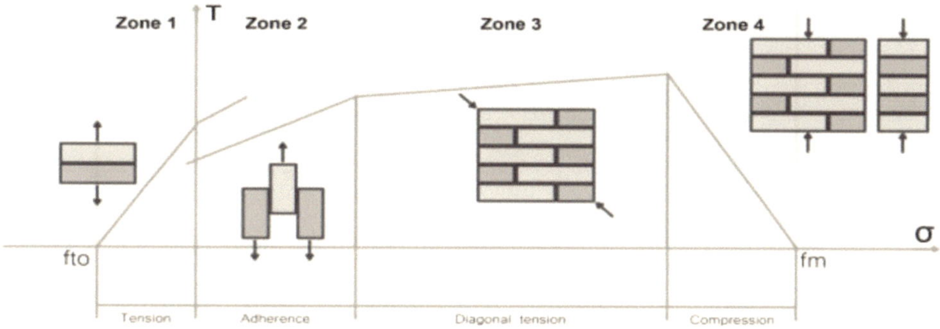

Figure 6.
Types of tests in masonry specimens made with blocks adhered with mortar a) Zone 1 – Tension, b) Zone 2 – Court, c) Zone 3 – Diagonal Compression, d) Zone 4 – Vertical Compression [37].

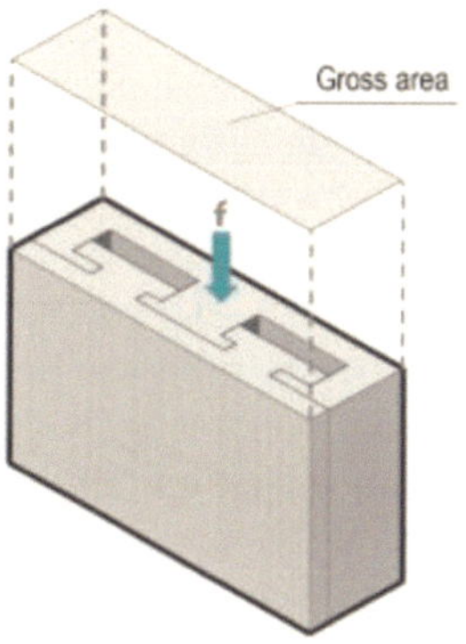

Figure 7.
Diagram of the simple compressive strength test for interlocking concrete blocks.

(temperature and humidity), the units were demolded (**Figure 7**) to be subjected to compression tests after 28 days of hardening.

3.2 Compressive strength test of the proposed system

The characterization of the mechanical performance of masonry elements made with blocks adhered by mortar bed is made from four tests: adhesion between blocks

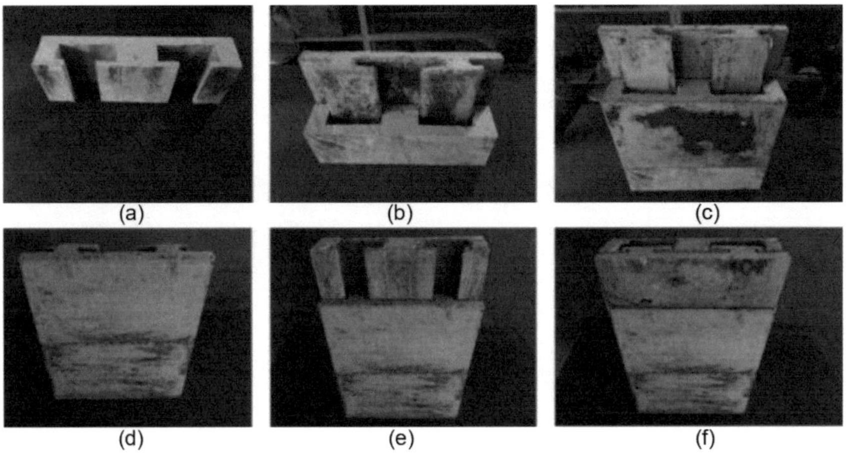

Figure 8.
Joint or manufacture of the prisms of interlaced blocks.

by traction and shear, resistance to a vertical compression fracture in block prisms and diagonal compression masonry assemblies (**Figure** 7) [38]. In this case, due to the type of union of the designed blocks and the construction process of the masonry, only the mechanical performance was considered from the last two tests, that is, axial compressive strength in prisms and diagonal compressive strength in masonry assemblies (**Figure** 8).

3.3 Compressive strength test of interlocking blocks

The compressive strength tests of the interlocking blocks were based on the NMX-C-404-ONNCCE standard. The NMX C404 is a Mexican standard that evaluates masonry pieces for structural use, describing dimensions, shape, test methods, classification, specifications in the way of placing the pieces and the values of compressive strength by type of piece [36]. Five blocks made with the M3 mixture (polypropylene sand, cement, water, fluidizer, and fiber) reported in **Table 2** were tested applying the load between the upper and lower face of each one of the blocks. In all cases, the specimens that did not present visible cracks and with good parallelism between their upper and lower faces were chosen, making sure to align vertically, horizontally, and the center of the block with the steel plate of the testing machine (ELE, 36-3088/02, series 040700000005), at a loading rate of 180 kg/cm2/min, according to the standard [36].

The compression of the blocks was obtained by dividing the fully factored load recorded by the total cross-sectional area of the sample (gross area) of a perpendicular section to the direction of the load, without discounting the gap (**Figure** 7), using the Formula 1, established in the NMX-C-036-ONNCCE standard [39].

$$f_p = P/A \tag{1}$$

Where f_p is the compressive strength, P is the total factored load supported by the block, and A is cross section of the gross or total area of the specimen.

Nevertheless, for design processes and calculations, the compressive design resistance (f_p^*) must be obtained using Formula 2, established by the NMX-C-404-ONNCCE standard [36].

Test	Norm	Sample	Area	Maximum load	Individual compressive strength (fp)	Medium compressive strength (f*p)
			cm	kgf	kg/cm²	kg/cm²
Compressive strength of pieces of block of concrete	NMX-C-404-ONNCCE-2012	1	496.17	92130	186.68	177.72
		2	498.75	88780	178.01	
		3	498.33	85720	172.01	
		4	497.5	118100	237.39	
		5	456.5	118100	223.79	
				Promedio:	**193.27**	

Table 3.
Results obtained from the compressive strength of concrete block pieces.

$$f_p^* = f_p / (1 + 2.5\, C_p) \tag{2}$$

Where f_p is the average compressive strength of five pieces using Formula 1, and *Cp* is a coefficient of variation of the compressive strength of the pieces, established in the NMX-C-404-ONNCCE standard, which will be taken as 0.35 for manufactured pieces or artisanal production and that do not have a quality control system, 0.30 for machining plants that do not have a quality control system, 0.20 for machining plants that demonstrate having a quality control system [36]. The results are shown in **Table 3**.

3.4 Fabrication of interlocked concrete block masonry and prism assemblies

For the compression test of prism and masonry assemblies made by joining and mechanical joining of the designed blocks, a total of three panels thick were made with specific dimensions, 40 centimeters in long, 62.5 centimeters stop, and 12.5 centimeters of thickness, in the case of the prisms (**Figure 9a**), and three panels with dimensions of 60 centimeters in length, 62.5 centimeters of stop, and 12.5 centimeters of thickness (**Figure 9b**), for the case of masonry assemblages, following that established in NMX-C-464-ONNCCE standard (**Figure 10**) [40].

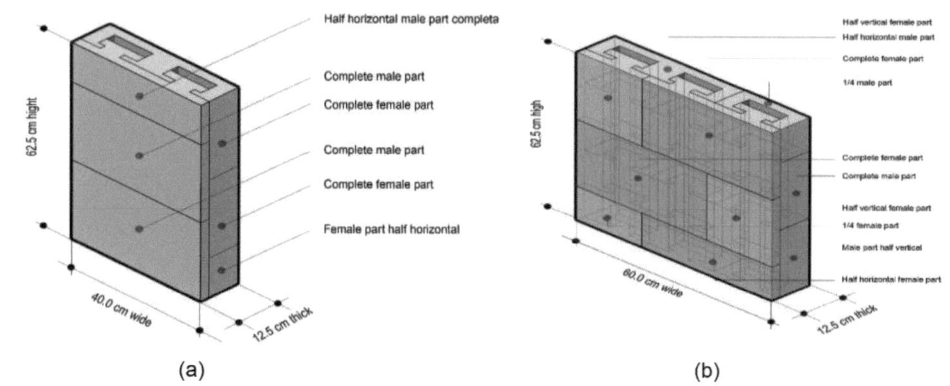

Figure 9.
Diagram of armed and dimensions of prisms and masonry assemblies made with interlocked mortar blocks.

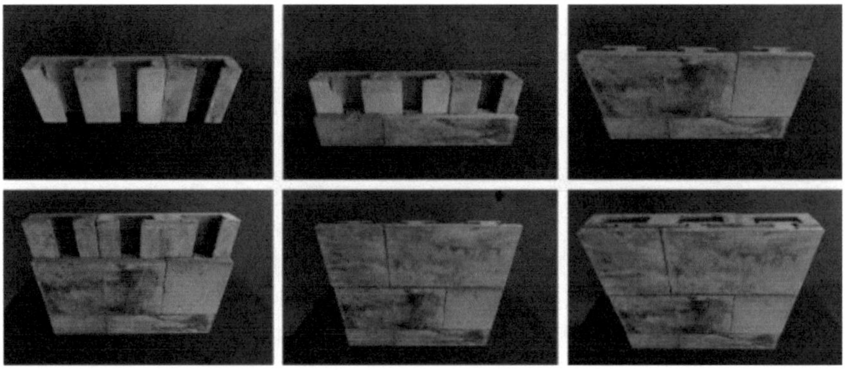

Figure 10.
Joint or manufacture of the masonry assemblages of interlaced blocks.

3.5 Axial compressive strength test of interlocking concrete block prisms

The determination of compressive strength of masonry was by testing three prisms of the same dimensions, which are built with the same type of pieces and technique. For this test, an ELE compression machine (model 1987AFE-X1) was used. This machine has a cushion made of steel plates. The female-male interlocking concrete blocks were placed edge to the steel plate with a thickness equal or greater than a third of the distance of the load-bearing block to the farthest corner of the sample, guaranteeing uniform distribution of the load according to the standard used [40]. The loading rate was 1.5 $kg/cm^2/seg$ [40].

The compressive strength of the prism was calculated according with the stipulated in NMX-C-464-ONNCCE standard [40], dividing the total factored load supported during the test by the gross load area of the prism (**Figure 11**), determined as an average at least three of the pieces of the prism (Formula 3). The result is expressed to an approximation of 0.01 MPa (0,1 kg/cm^2), the resistance obtained is multiplied by the correction factor for slenderness indicated in **Table 4** [40].

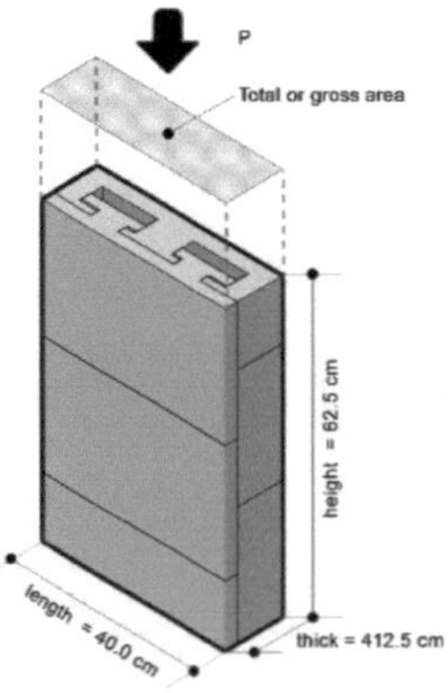

Figure 11.
Diagram of compressive strength in prisms made with blocks of concrete interlaces.

Test	Norm	Sample	Large de la pila (mm; cm)	Prism thickness (mm; cm)	prism height (mm; cm)	Area	Slenderness ratio	Correction factor	Maximum load (N; kgf)	Compressive strength (kg/cm^2)	Design Compressive strength (kg/cm^2)
Compressive strength concrete block piles	NMX-C-464-ONNCCE-2012	1	40	12.5	62.5	500	5.00	1.0521	30972.9	65.17	47.4
		2	40	12.5	62.5	500	5	1.0521	27498.3.	57.86	
		3	40	12.5	62.5	500	5	1.0521	25939.3	54.58	

Table 4.
Results obtained from the compressive strength in prisms made with concrete block of interlaces.

Stack slenderness ratio	Corrective factor
2	0.750
3	0.900
4	1.000
5	1.050
6	1.060

Table 5.
Correction factors for slenderness of the prisms obtained from [40].

$$f_m = P/tb \; x \; slenderness \; factor \tag{3}$$

Where f_m is the compressive strength of the prism in MPa (kg/cm^2), P is the total applied load in N (kg), t is the thickness of the prisms in mm (cm), and b is the width of the prisms in mm (cm), the slenderness factor is indicated in **Table 5**.

The design compressive strength is calculated with Formula 4, established in the standard [40]:

$$f_m^* = f_m/1 + 2.5C_m \tag{4}$$

Where f_m^* is the compressive strength for design purposes in MPa (kg/cm^2), f_m is the average of the resistant efforts of the tested prisms; referred to the gross area give MPa (kg/cm^2), and *Cm* is the coefficient of variation of the resistant efforts of the tested prisms, calculated as the quotient of the standard deviation between the average, and that should not be taken less than 0.10 in the case of verifying the quality control in work, nor that 0.15 in the other cases, according to what is established in the Mexican standard: NMX-C-464-ONNCCE [40]. The results are summarized in **Table 4**.

3.6 Diagonal compressive strength on the panel made with interlocking concrete blocks

Figure 12 shows the diagonal compressive strength test system for masonry assembly. For this test, a load was applied along with one of the diagonals of the specimen. This process is established by NMX-C-464-ONNCCE [40]. Briefly, before the total load, masonry assemblies are carefully aligned to the axis of the machine with the axis of the sample. According to the same standard [40], three cycles of preload with 15% of the total load should be applied (20 kg/cm^2). Finally, the loading rate was of 1.5 kg/cm^2/seg [40]. These tests were done in an ELE compression testing machine (model 1987AFE-X1). Three masonry assemblies were tested, as established by the NMX-C-464-ONNCCE standard [40].

The diagonal compression resistance was obtained using Formula 5 established in the NMX-C-464-ONNCCE standard [40], which defines the resistance of each wall as the ratio between the total factored load and the gross area of the masonry assemblies. The latter is obtained from the product of the thickness of the masonry assemblies (t) by the length of the compression diagonal (Lc) (**Figures 13** and **14**) measured before the test.

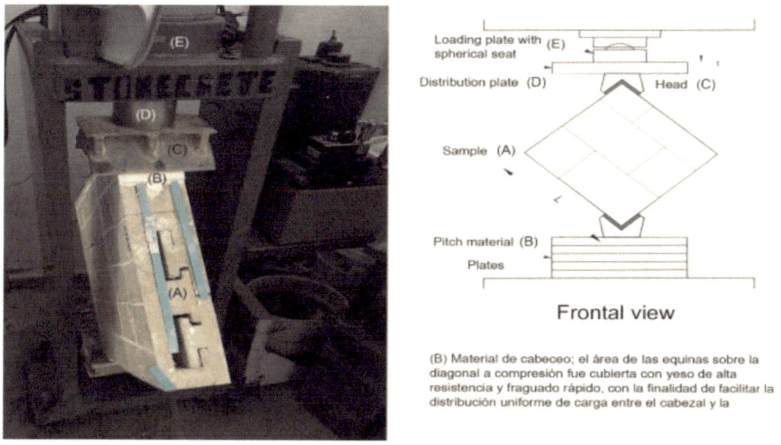

Figure 12.
Diagonal compressive strength test system for masonry assemblies made with interlocking concrete blocks, obtained, and adapted from the NMX-C-464-ONNCCE standard [40].

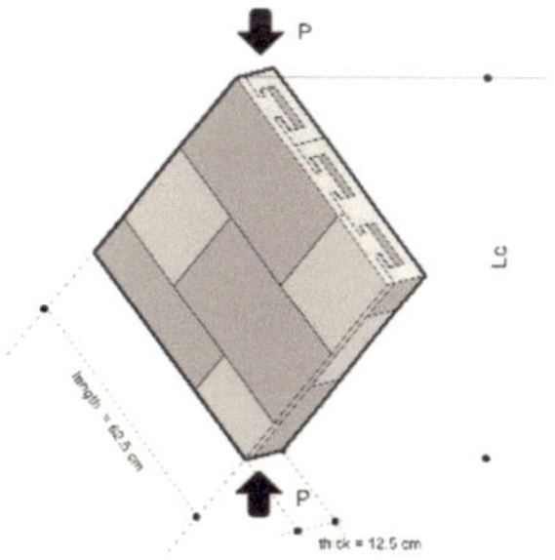

Figure 13.
Diagonal compressive strength diagram for masonry assemblies made with interlocking concrete blocks.

a) b)

Figure 14.
a) Female-male concrete block after the compressive strength test, b) typical cracking on the lateral faces of the female-male concrete blocks after the compressive strength test.

$$v_m = P/tL_c \tag{5}$$

Where *Vm* is the diagonal compressive strength of the masonry assemblies in MPa (kg/cm^2), *P* is the total applied load in N (kgf), *t* is the thickness of the masonry assemblies in mm (cm), and *Lc* is the length of the compression diagonal in mm (cm).

The compression resistance due to diagonal traction for design purposes is calculated using Formula 6 established in Standard NMX-C-464-ONNCC [40]:

$$v_m^* = v_m/1 + 2.5C_v \tag{6}$$

Where v_m^* is diagonal compressive strength for design purposes in MPa (kg/cm^2), *Vm* is the average of the resistant stresses of the tested masonry assemblies referred to the gross area in MPa (kg/cm^2), *Cv* is the coefficient of variation of the resistant efforts of masonry assemblies tested, calculated as the quotient of the standard deviation between the average and which should not be taken less than 0.10 in the case of verifying the quality control on site, nor than 0.20 in other cases [40], obtaining the results in **Table 6**.

4. Discussion of results

4.1 Mortar mixtures

High-performance mortars must have the following characteristics [41–44]:

1. In their fresh state, they must be able to flow with good adhesion, to be placed without segregation or stratification, and must have an excellent performance on construction sites.

2. Hard mortars must have high mechanical resistance to compression.

3. In the hardened state, they must have high volume stability, i.e., low shrinkage and low warpage.

4. They must have discharge durability.

To obtain these benefits, new technologies have been developed: On the one hand, a meticulous selection of the particle size of the high-quality stone aggregates achieving the adequate packing of these elements in the prepared mixtures. The uses of additives allow to adjust the physical-chemical properties of hydrated calcium silicates (CSH) that constitute the binder that provides the mechanical properties to concrete and mortar [45].

Figure 4 shows the effect that the additives have on the compression fracture of the mixture M3 (plasticizer and polypropylene microfibers (**Figure 4c**), concerning the mixtures M1 (**Figure 4a**) and M2 (**Figure 4b**). In the M1 mortars, used as a reference, made with sand, cement, and water, the almost total disintegration of the material is observed when the fracture point is reached. In the remaining cases, samples M2 and M3, the disintegration is partial. In the case of M2 samples (mixture with fluidizer), the shape of the piece is maintained, and multiple cracks only appear at the moment of rupture. In the case of M3 samples (mixture with fluidizer and

DOI: http://dx.doi.org/10.5772/intechopen.107174

Test	Norm	Sample	Maximum load (N; kgf)	Die thickness (mm; cm)	Diagonal length (mm; cm)	Diagonal compressive strength (kg/cm^2)	Diagonal compressive strength (kg/cm^2)
Diagonal compression of dies manufactured with tongue-and-groove block	NMX-C-404-ONNCCE-2012	1	5233.7	12	86	**5.07**	**3.98**
		2	4835.3	12	87	**4.63**	
		3	5445.7	12	87	**5.22**	

Table 6.
Results obtained from the compressive strength in masonry assemblages diagonal made with concrete block of interlaces.

polypropylene microfibers), the piece integrity is almost total with some cracks. These results clearly show the effect of fluidizers and polypropylene microfibers on the structural integrity of mortars. The effect of these additives is also observed in the mechanical performance of the mixtures, with the maximum breaking strength being 210.57 kg/cm^2 for the mixture M1. In the case of mixtures M2 and M3, these values were 375.64 and 341.64 M3 kg/cm^2, respectively, showing an increase of >60% of the mechanical resistance when the additives are added, this is mainly due to the reduction of the water-cement ratio [46]. As a reference and for the case of Mexico City, fracture resistance values in compression of a type 1 structural mortar are 180 kg/cm^2, with mixture M1 and mixture M2 between 180% and 200% above the accepted value by this standard [37]. Although the mechanical resistance to fracture of the M2 mix is higher than that of the M3 mix (around 9% less than M3), the structural integrity observed in the latter (**Figure 4c**) makes it a candidate to be used in masonry constructions in areas of high seismicity. Moreover, there is the greatest possible ductility so that the structure can dissipate the greatest amount of energy in the event of an earthquake.

4.2 Concrete blocks

As already mentioned, M3 mix was chosen to build the interlocking concrete blocks, under the hypothesis that resilient masonry systems with good mechanical performance could be made with them, manufacturing blocks with dimensions of 40 cm long x 25 cm high x 12,5 cm wide and taking them to compressive strength tests after 28 days, and for both pieces, the wall, and the prism systems. **Table** 7 compares the design resistance to compression of the blocks, masonry assemblies, and prisms made, referring to the compressive strength requirements of common masonry with concrete blocks and cement mortar, required by the Complementary Practical Standards of Masonry of Mexico City.

The results of **Table** 7 show that the average compressive strength of the interlocking blocks exceeds the established on the NMX-C-036-ONNCCE standard (150 kg/cm^2) by almost 30 kg/cm^2, although for some individual cases from these blocks, values close to 200 kg/cm^2 were obtained, i.e., 50 kg/cm^2 above what was established. This indicates that it is possible to increase the mechanical performance of the blocks made with the M3 mixture.

Another advantage of manufacturing the blocks with this type of mixture is that the fracture process did not lead to high fragmentation. This is because the polypropylene fibers act as a three-dimensional reinforcement in the mortar since they help to

Test	Design strength (kg/cm^2)	Design strength (kg/cm^2)	
	NORMATIVE	PROYECT	
Block	150	180	✓
Prims	25–100	47.4	✓
Assemblage	2.0	3.98	✓

Table 7.
Comparison of results obtained from the compressive design resistance of blocks, prisms, and interlocking concrete block masonry assemblies vs. the resistance established in the NMX-C-464-ONNCCE and NMX-C-036-ONNCCE standards for pieces, prisms, and masonry concrete block assemblies.

distribute internal tensile stresses and flexion of homogeneous compression and bending during compression tests, in addition to reducing microcracks and cracks induced by temperature changes and plastic contraction during the mortar setting process during block manufacture.

The failure of the fiber-reinforced blocks occurs gradually, and the cracks derived from the test are visible extending through the lateral faces of the specimen. The fibers allow the blocks to resist a small increase in load after cracking and increase the toughness of the units.

On the other hand, and as we have already seen, the compressive strength in prisms made with interlocking blocks was 47.4 kg/cm^2, higher than the average design strength required by the standard for prisms made with higher strength concrete blocks at 60 kg/cm^2 together with type 1 mortar that establishes a resistance of 25 kg/cm^2, and lower prisms made with blocks with a resistance of 150 kg/cm^2 and joined with type 1 mortar where the required resistance is 75 kg/cm^2, but very close prisms made with blocks with a resistance of 100 kg/cm^2 and type 1 mortar, where the minimum design resistance required by the standard is 50 kg/cm^2.

During the application of the load, cracks were originated in the front faces of the prisms (long side in the female type pieces). Failure usually occurs from compression cracking or vertical cracking. This type of failure is produced by the difference in the deformable cross section between the pieces, generating tensile stresses in the latter. The fibers added to the concrete blocks help to withstand the stresses, thereby reducing and controlling the propagation of crack opening. As there are no lines of fragility as it is with a prism where the mortar intervenes suddenly, the concrete blocks work at their maximum capacity, generating a longer rupture or failure time compared with a prim adhered with mortar.

On the other hand, in the design resistance to diagonal compression of the masonry assembly, an average load of 3.9 kg/cm^2 was obtained, complying with the requirement established by the Mexican standard, which is 2 kg/cm^2 for solid pieces with the structural application; however, in the individual behavior of masonry assemblies, the highest resistances were obtained at 5 kg/cm^2.

According to [47], masonry assemblies are the specimens that best capture the failure modes of structural masonry, since they consist of blocks joined by mortar through the aligned, continuous, or horizontal bed, and stepped vertical bed. This arrangement allows the polypropylene fibers to exhibit their explosion effect in the best way, since, although there is no encounter with the mortar, the stresses developed in the units are greater than in the blocks and prisms. This is basically due to two factors: first, the larger sample size, being less bound by the test device than the other samples. Second, the presence of dry or staggered butt joints results in higher stresses compared with prisms. Under these conditions, the fibers act as an effective reinforcement and can effectively contribute to improving the strength of the structural element.

In the test, the vertical load generates increasing tensile stresses that are oriented perpendicular to the load direction. This tensile stress field leads to the failure of masonry assemblies along a crack approximately perpendicular to the diagonal between the two loaded corners.

During the tests carried out on the masonry assemblies, it was observed that the fault was combined, since the pieces slipped due to lack of adhesion between them; however, when the pieces reached their maximum comfort of the displacement due to the system, the force that interacts on the overlaps between the blocks causes them to work as if it were normal masonry, generating diagonal tensions where the cracks cross the pieces indistinctly. On the other hand, a form of ductile failure occurs since the adherence of the fiber-matrix allows the parts of the cracked elements to remain united, generating advantages for structural systems in seismic zones by having more ductile elements that give time to react to a total collapse. According to Mehta and Monteiro [46], fiber-reinforced concrete will suffer increasing loads after the first cracking of the matrix due to the resistance to fiber extraction. As the load increases, the fibers tend to transfer the additional stress to the surrounding matrix through the bonding stresses until fiber failure occurs or until the accumulated slip locally leads to fiber breakage.

5. Conclusions

In this study, an interlocking procedure of blocks made with structural class mortars with high mechanical resistance to compression is proposed as a construction method according to the Mexican standard. The resistances obtained in the experimentation of the compression of prisms (axial compression) and masonry assemblies (diagonal compression) made with those blocks comply with the values required by the standard for common masonry systems of ordinary structural blocks adhered to common concrete with hitting mortar. Therefore, the system developed in this research can be used in compression forces as an alternative to the conventional construction system, with the additional advantage that it represents the simplicity of the assembly for the manufacture of mechanical elements, the physical problem of the walls, and the reduction of and manufacturing between the stick mortar and the masonry, accelerating the production and reducing costs, as it does not require specialized equipment or labor.

The results obtained in the experimental process suggest that the construction method with load-bearing walls made with the type of blocks developed in this research can generate more efficient housing construction methods. Nevertheless, before establishing general conclusions about the mechanical behavior of masonry made with these blocks, additional studies and exhaustive tests related to the elastic constants of the wall sections must be carried out, to establish design criteria (failure limit state, serviceability limit state), design for durability, resistance factors, evaluate masonry walls through their confinement or interior reinforcement, to have elements that allow the foundation of structural calculation principles and ensure the performance of the masonry units or pieces in the present structural masonry systems, complying with the structural mechanical performance assumptions, such as the capacity for deformation and ductility, to guarantee the stability of a structure built with this system, according to normative annex A, "number of acceptance of defects of construction systems with masonry designed for earthquakes," established in the complementary practical standards for the masonry of Mexico City. However, the acceptance or rejection of the new system will be the responsibility of the competent authority of Mexico City, which is the Construction Safety Institute, this unit will assess if the system complies with the current standard in force under the scientific, technical, and technological aspects.

Finally, a stress concentration analysis must be carried out to optimize the geometry of the interlocking blocks to avoid mechanical failure in weak areas, as well as generate a mixture of semi-wet or dry mortar to be able to optimize their industrial production.

Author contribution

Edrey Nassier Salgado Cruz: Writing – Reviewing, Software, Investigation. Alberto Muciño: Writing – Reviewing, Conceptualization, Methodology. Eligio Orozco: Supervision, Writing-Reviewing and Editing, Investigation. César Armando Guillén Guillén - Writing-Reviewing and Editing, Investigation

Acknowledgements

This work was supported by the Dirección General de Asuntos del Personal Académico (DGAPA-UNAM) under contract PAPIIT-IN101221.

Author details

Edrey Nassier Salgado Cruz[1], Alberto Muciño Vélez[2]*,
Eligio Alberto Orozco Mendoza[3] and César Armando Guillén Guillén[1]

1 Faculty of Architecture, National Autonomous University of Mexico, S/N School Circuit, University City, Coyoacán, Mexico City, Mexico

2 Center for Research in Architecture, Urbanism and Landscape CIAUP, S/N School Circuit, Faculty of Architecture, National Autonomous University of Mexico, University City, Coyoacán, Mexico City, Mexico

3 Institute of Physics, Investigation Circuit S/N, National Autonomous University of Mexico, University City, Coyoacán, Mexico City, Mexico

*Address all correspondence to: amucino@unam.mx

References

[1] Guillaud H. Characterization of Earthen Materials. Terra Literature Review—An Overview of Research in Earthen Architecture Conservation. 2008. p. 21. Available from: https://getty.edu/conservation/publications_resources/pdf_publications/pdf/terra_lit_review.pdf#page=34

[2] Waleed A, Mohd-Saleh J, Mohd-Razali A, Kadir A, Abdullah A, Trikha D, et al. Development of an innovative interlocking load bearing hollow block system in Malaysia. Construction and Building Materials. 2004;**18**:445-454

[3] Tingwei S, Xihong Z, Hong H, Chong C. Experimental and numerical investigation on the compressive properties of interlocking blocks. Engineering Structures. 2021;**228**: 111561

[4] Rekik A, Allaoui S, Gasser A, Blond E, Andreev K, Sinnema S. Experiments and nonlinear homogenization sustaining mean-field theories for refractory mortarless masonry: The classical secant procedure and its improved variants. European Journal of Mechanics - A/Solids. 2015;**49**:67-81

[5] Majid A, Romain B, Nawawi C. Dynamic response of mortar-free interlocking structures. Construction and Building Materials. 2013;**42**:168-189

[6] Martínez M, Atamturktur S. Experimental and numerical evaluation of reinforced dry-stacked concrete masonry walls. Journal of Building Engineering. 2019;**22**:181-191

[7] Anand K, Vasudevan V, Ramamurthy K. Water permeability assessment of alternative masonry systems. Building and Environment. 2003;**38**(7):947-957

[8] Ramamurthy K. Accelerated masonry construction review and future prospects. Advances in Structural and Materials Engineering. 2004;**6**(1):1-9

[9] Sokairge H, Rashad A, Elshafie H. Behavior of post-tensioned dry-stack interlocking masonry walls under out of plane loading. Construction and Building Materials. 2017;**133**:348-357

[10] Al-Fakih A, Wahab M, Mohammed BS, Liew MS, Zawawi WA, As'ad S. Experimental study on axial compressive behavior of rubberized interlocking masonry walls. Journal of Building Engineering. 2020;**29**:101107

[11] Dyskin AV, Pasternak E, Estrin Y. Estructuras sin mortero basadas en enclavamiento topológico. Frente. Estructura civ. Ing. 2012;**6**:188-197

[12] Hydraform. Hydraform Training Manual. Johannesburg 220 Rondebult road, Libradene boksburt. P.O. Box 17570, Sunward park 1470. www.hydraform.com. 2004

[13] Anand KB, Ramamurthy K. Laboratory-based productivity study on alternative masonry systems. ASCE Journal of Construction Engineering and Management. 2003;**129**(3):237-242

[14] Grimm CT. Masonry construction operations. Journal of the Construction Divisions. 1974;**100**(2):71-185

[15] Sadafi N, Zain MFM, Jamil M. Structural and funtional análisis of an industrial, flexible, nd demountable wall panel system. IJE. 2014;**27**(2):247-260

[16] Anand, K.B. and K. Ramamurthy, Techniques for accelerating masonry construction. International Journal for

Housing Science and Its Applications. 1999:.23 (4):.233-241. Available in: https://www.researchgate.net/profile/Ramamurthy-K/publication/292668508_Techniques_for_accelerating_masonry_construction/links/5d86703ea6fdcc8fd60bf0dd/Techniques-for-accelerating-masonry-construction.pdf

[17] Mirasa, AK, Chong, CS. La construcción de edificios ecológicos mediante el sistema de ladrillos entrelazados. En: Yaser, A. (eds) Green Engineering for Campus Sustainability. Springer: Singapur. 2020. P. 35-49.

[18] Fundi SI, Kaluli JW, Kinuthia J. Behavior of interlocking laterite soil block walls under static loading. Construction and Building Materials. 2018;**171**:75-82

[19] Javan AR, Seifi H, Xu S, Xie Y. Design of a new type of interlocking brick and evaluation of its dynamic performance. In: Proceedings of IASS Annual Symposi. Vol. 2016, No. 7. Athens, Greece: International Association for Shell and Spatial Structures (IASS); 2016. pp. 1-8. Available in: https://www.ingentaconnect.com/content/iass/piass/2016/00002016/00000007/art00005

[20] Ayed HB, Limam O, Aidi M, Jelidi A. Experimental and numerical study of Interlocking Stabilized Earth Blocks mechanical behavior. Journal of Building Engineering. 2016;7:207-216

[21] Allaoui S, Rekik A, Gasser A, Blond E, Andreev K. Digital Image Correlation measurements of mortarless joint closure in refractory masonries. Construction and Building Materials. 2018;**162**:334-344

[22] Fukumoto Y, Yoshida J, Sakaguchi H, Murakami A. The effects of block shape on the seismic behavior of dry-stone masonry retaining walls: A numerical investigation by discrete element modeling. Soils and Foundations. 2014;**54**(6):1117-1126

[23] Okail H, Abdelrahman A, Abdelkhalik A, Metwaly M. Experimental and analytical investigation of the lateral load response of confined masonry walls. HBRC Journal. 2016;**12**(1):33-46 https://www.tandfonline.com/action/showCitFormats?doi=10.1016/j.hbrcj.2014.09.004

[24] Mubeena S, Krishana SR. Seismic analysis of interlocking bloques in walls, IJSRD - International Journal for Scientific Research & Development. 2018:**6**(5):2321-0613. Available in: https://www.ijsrd.com/articles/IJSRDV6I50234.pdf

[25] Tyas, I. W. (2018). Lock-brick system for sustainable and environment infrastructure building materials. In IOP Conference Series: Materials Science and Engineering. 2018:**371**(1):012016. Available in: https://iopscience.iop.org/article/10.1088/1757-899X/371/1/012016/meta

[26] Al-Fakih A, Mohammed BS, Liew MS, Nikbakht E. Incorporation of waste materials in the manufacture of masonry bricks: An update review. Journal of Building Engineering. 2019;**21**:37-54

[27] Tena A, Miranda E. Capítulo 4: Comportamiento mecánico de la mampostería. Edificaciones de Mampostería para la Vivienda. 2004;**5520**(2):103-132

[28] Susilawati, CL, Suni, PY y Tjandra, E. (2020). Lock-brick system technology is an ecological building material innovation, Earth and Environmental

Science. 2020:**419**(1):012005. https://iopscience.iop.org/article/10.1088/1755-1315/419/1/012005/meta

[29] Tena A, Liga A, Pérez A, González F. Propuesta de mejora de mezclas para producir piezas de mampostería de concreto empleando materiales comúnmente disponibles en el Valle de México. Revista ALCONPAT. 2017;7(1):36-56

[30] ASTM. C136 / C136M-19, Standard Test Method for Sieve Analysis of Fine and Coarse Aggregates. West Conshohocken, PA: ASTM International; 2019

[31] ASTM D 75-97 AMERICAN SOCIETY FOR TESTING AND MATERIALS 100 Barr Harbor Drive, West Conshohocken, PA 19428 Reimpresa del Anuario de las Normas ASTM. Copyright ASTM

[32] Muciño A, Vargas S, Pérez NA, Bucio G, Orozco E. The influence of fine aggregates on Portland cement mortar compressive strength. Results in Materials. 2021;**10**:100182

[33] Bisby LA, Kodur VK. Evaluating the fire endurance of concrete slabs reinforced with FRP bars: Considerations for a holistic approach. Composites Part B: Engineering. 2007;**38**(5–6):547-558

[34] ASTM C109/C109M-20. Standard Test Method for Compressive Strength of Hydraulic Cement Mortars (Using 2-in. or [50-mm] Cube Specimens). 2020

[35] NMX-C-038-ONNCCE-2013. Building industry – Masonry – Determination of dimensions of blocks and bricks – Test method. 2013

[36] NMX-C-404-ONNCCE-2012. Construction industry - masonry – blocks or bricks for structural use - specifications and test methods (in Spanish). 2012

[37] NTC. Complementary technical standards for design and construction of masonry structures (in Spanish). 2020

[38] Mann W, Muller H. Failure of shear stressed masonry an enlarged theory, tests, and application to shear walls, Proceedings of the British Ceramic Society. 1982: 30:223-235. Available in: http://pascal-francis.inist.fr/vibad/index.php?action=getRecordDetail&idt=PASCALBTP83X0254419

[39] NMX-C-036-ONNCCE-2013. Building industry – Masonry – Compressive strength of blocks, bricks and pavers – Test method. 2013. p. 10

[40] NMX-C-464-ONNCCE-2010. Building industry – Masonry – Determination of diagonal compression strength and wall shear modulus, as well as determination of compressive strength and elastic modulus of masonry concrete and clay piles – Test method. 2010. p. 24

[41] NMX-C-414-ONNCCE-2017. Building industry – Hydraulic cementants – specifications and tests methods (in Spanish). 2017

[42] Bansal PP, Sharma R. Development of high-performance hybrid fiber reinforced concrete using different fine aggregates. Advances in Concrete Construction. 2021;**11**(1):19-32

[43] Malier Y. High Performance Concrete (from Material to Structure). London: E & FN SPON; 1992. p. 270

[44] Ozawa K, Maekawa K y Okamura H. Development of High-Performance Concrete, University of Tokyo, Japan. 1992. Available in: https://www.osti.gov/etdeweb/biblio/6509280

[45] Falikman V. Defined-Performance concretes using nanomaterials and nanotechnologies. Engineering Process. 2022;**17**:12

[46] Mehta PK, Monteiro PJM. Concrete: Microstructure, Properties, and Materials. 4th ed. California, United States of America: McGraw-Hill; 2014

[47] Hendry AW. Masonry walls: materials and construction. Construction and Building Materials. 2001;**15**(8):323-330

Chapter 5

The Strength of Masonry Based on the Deformation Characteristics of Its Components

Alexey N. Plotnikov, Viktor A. Ivanov, Boris V. Mikhailov, Tatyana G. Rytova, Olga S. Yakovleva, Mikhail Yu Ivanov and Natalia V. Ivanova

Abstract

The chapter presents a new approach to determining the strength of masonry reinforced with transverse meshes in mortar joints. The method consists of using the values of the modulus of elasticity and limiting deformations of the stone material, mortar for joints, and both steel and composite reinforcements. An analytical notation is proposed that integrally takes into account the characteristics of the initial materials. The results of physical tests of centrally loaded masonry pillars reinforced with steel and composite meshes are given. To test the masonry, widely used materials were used: solid brick and cement-sand mortar. The values of the bearing capacity, deformations, and internal stresses of the masonry are obtained. It is determined that the stresses in the reinforcing bars of the meshes are unevenly distributed in the horizontal plane of the mortar joint and amount to 20–37% of the design resistance of the mesh material. The strength of masonry reinforced with composite meshes is 65–75% of steel of the same cross section. It is shown that there is a good convergence of test results with the presented analytical dependence.

Keywords: masonry, reinforcement, deformation, strength, testing, modulus of elasticity, composite reinforcement, steel reinforcement, basalt bars, reinforcement mesh, design, the percentage of reinforcement

1. Introduction

In construction practice, the method of increasing the bearing capacity of masonry using mesh reinforcement in horizontal mortar joints when working in central compression is quite widespread. Recently, along with a metal mesh, meshes made of composite reinforcement (fiberglass, basalt plastic, and others) have been used. The physical essence of the method is to contain the transverse deformations of the masonry and transfer the part of the forces to the reinforcing bars located in the horizontal mortar joints. The calculation of such masonry is regulated by the set of rules SP 15.13330.2020 "Stone and reinforced masonry structures." According to this standard, masonry is considered a homogeneous structure, while the given physical and mechanical characteristics of its components (brick and mortar) are used. The same concept underlies Eurocode 6. The second approach is to represent masonry as a complex composite structure with materials of different modulus having significantly different characteristics.

In any case, like concrete, masonry is reinforced to give a brittle material—stone, which has high compressive strength and greater tensile strength. Reinforcement makes it possible to increase the strength of the stone by preventing lateral expansion caused by a force applied perpendicular to the mesh. According to the standard (set of rules) SP 15.13330.2020, the strength of reinforced masonry doubles.

In fact, reinforced masonry is a composite material consisting of the main mass in the form of stone, interlayers of mortar, and rarely ordered inclusions in the form of steel or composite rods. The use of materials with different characteristics requires the creation of calculation methods that take into account their initial characteristics. Appropriate diagrams of material deformation are needed—in tension, shear, and compression.

According to the works of V.A. Ivanov, L.I. Vucin, M.V. Skobeeva, A.I. Kibets, Yu. I. Kibets [1, 2] in a material where the binder matrix has numerous differently directed more rigid inclusions, it is difficult to establish the actual distribution of strains and stresses.

In the work of S. Babaeidarabad [3], it is established that in order to increase the strength when strengthening the masonry, there is a ratio of parameters, in particular, reinforcement coefficients. However, in this work, only external reinforcement is discussed.

The continuum model of masonry without reinforcement has its place, especially when analyzing the nonlinear behavior of a structure. Models, according to A.H. Akhaveissy, can take into account microcracks in masonry, which lead to softening and destruction [4].

In any case, researchers proceed from the definition of the parameters that make up the masonry, using them for either discrete or continual model building. For masonry, predictive analytical dependencies can be obtained to build nonlinear graphs of masonry work. In the work of T.C. Nwofor [5], obtained nonlinear tension curves with characteristic points highlighted the tension curves with a stress level of 0.4 from the breaking load, which corresponds to the limit of the near linear region.

Works of V.A. Ivanov, L.I. Vucin, M.V. Skobeeva, A.I. Kibets, Yu. I. Kibets [1, 2] showed that in this case, brickwork can be modeled as a continuum multimodular medium, the properties of which depend on the type of stress-strain state and the current level of damage to the material. To calculate masonry, a simplified model can be applied that takes into account the deformability of joints, the strength of brick, and mortar in tension and shear, as well as the contact interaction of masonry fragments. Each brick is divided into a number of segments (blocks). The brick material is assumed to be isotropic and ideally elastic. The destruction of masonry along horizontal and vertical seams and along sections of bricks that bind vertical seams is considered. At the initial stage (before destruction), when analyzing the interaction of two blocks of one brick, the contact pressure components are calculated from the conditions of rigid gluing. The stresses in the joints of the masonry are determined through the deformation of the binder.

With a contact pressure component $q_n > 0$, the tensile and shear strength criteria are checked in succession. With compression ($q_n < 0$), only the fulfillment of the criterion for shift is analyzed. If at least one of the strength criteria is violated, it is considered that local destruction of the brickwork has occurred, and in the future, the contact interaction at this point is modeled using the friction algorithm.

Recently, more and more two described methods penetrate into each other and are used in a complex, as can be seen from a number of works.

Based on the theory of resistance of anisotropic materials, A.B. Antakov and B.S. Sokolov [6, 7] obtained masonry deformation diagrams under compression. Taking into account a large number of tests, the stages of the stress state were described. The values of the tear, shear, and crush forces were determined using the strength characteristics of the masonry: tensile strength R_t, shear R_{sh}, compression R, and geometric parameters—the areas of the corresponding surfaces A_t, A_{sh}, A_{ef}.

Techniques for modeling masonry by the finite element method are being developed with the introduction of a number of specific physical and mechanical parameters. G.G. Kashevarova [8] introduced criteria into the model that take into account orthotropy, strength, strain softening, and layer shear coefficients while ensuring a minimum level of resistance. The criteria for the strength of individual components are accepted: brick and mortar in tension and shear, and strength of contact between brick and mortar in a horizontal joint. The angles of inclination of the load and the ratio of types of load that affect the strength of the masonry are determined.

A deep analysis of the influence of masonry components, including the location of brick faces, in the form of finite element models with the inclusion of empirical data, was carried out by V.V. Pangaev [9]. This makes it possible to select the composition and system of bonding masonry while taking into account the different nature of deformation and destruction of typical elements of masonry—bonded and spoon rows, and vertical and horizontal mortar joints. It is shown that a small physical sample of five spoon rows of bricks is sufficient to obtain reliable data on the stress-strain state and can be accepted as a masonry element.

The complex model of O.V. Kabantsev [10] combines discreteness and has elements in the form of individual bricks and layers of mortar, and continuity, a material with properties that take into account the contact interaction of the constituent components of the masonry.

Most modern authors use piecewise homogeneous physically nonlinear functions of individual components to build models [1, 11–15].

Recently, the use of reinforcement in the form of meshes of composite rods in horizontal joints of masonry has been growing [16–23]. This increases the heat transfer resistance of the outer walls, increases the corrosion resistance of reinforcement, and, in some cases, reduces the cost of reinforcement. However, the use of composite reinforcement in masonry is constrained by the lack of calculation methods and experimental data.

2. Materials and research methods

The traditional method for calculating masonry reinforced with meshes, given in the design standards (SP 15.13330.2020), is based on empirical dependencies obtained by L.I. Onishchik [24, 25]. On the whole, it has justified itself for several decades of application for steel meshes, but it does not take into account the peculiarities of the physico-mechanical properties of composites at all. Composite, in particular, basalt-plastic-reinforced, as part of the structure, manifests itself as a very strong material in tension, having a tensile strength of at least R_f = 1000 MPa. However, the elastic modulus, in this case, is only E_f = 50,000 MPa [7]. For steel, this ratio is different (Rs = 400 MPa, Es = 200,000 MPa).

Transverse reinforcement in the form of meshes is used to increase the bearing capacity of the masonry in compression. According to current standards, the amount of reinforcement in the masonry is determined by the percentage of reinforcement by volume:

$$\mu = \frac{V_a}{V_k} 100, \tag{1}$$

where $V_a = (C_1 + C_2)A_{st}$—reinforcement volume, $V_k = C_1 C_2 S$—masonry volume, S—height spacing of grids.

The minimum percentage of reinforcement is assumed to be $\mu_{min} = 0.1\%$ and maximum $\mu_{max} = 1\%$.

The tensile strength of masonry with mesh reinforcement is determined by the formula:

$$R_{sku} = kR + \frac{2R_{sn}\mu}{100}, \tag{2}$$

where R_{sn} is the normative tensile strength of reinforcement; R is the tensile strength of the masonry; k is a coefficient that takes into account the type of stone.

For reinforcement made of class B500 steel, R_{sn} is taken with a reduction factor of working conditions of 0.6. Therefore, it is considered that the limit of resistance is not reached in the reinforcement during the destruction of the masonry. However, there are practically no experimental data confirming this norm.

The fracture mechanics of masonry assumes the occurrence of critical tensile stresses in the transverse direction of the vertical element under the influence of the Poisson effect at a stress level of 0.4–0.7 of Ru (tensile strength) depending on the ratio of strength and modulus of elasticity of stone and mortar. More often, vertical power cracks occur above vertical mortar joints, less often along the stone, when the mortar bed is not made evenly enough. The destruction occurs from the rupture of stones, and the masonry is divided into separate columns, a multiple of half the brick in size.

To increase strength and reduce deformations in the transverse direction of the masonry, reinforcement with metal or composites in the form of meshes in horizontal mortar joints is used [1, 2, 14, 16–23]. Part of the stress is transferred to the reinforcement. Cracking, in this case, is not so intense, and cracks appear at stress levels above 0.7 Ru. The division of masonry into separate columns does not occur. Therefore, the level of stress in the reinforcement is important for the calculations of reinforced masonry.

To consider the stresses in the volume of masonry, it is necessary to connect them with Gook-law (**Figure 1**):

$$\varepsilon_x = \frac{1}{E}\left[\sigma_x - \nu\left(\sigma_y + \sigma_z\right)\right] \tag{3}$$

The Poisson ratio for masonry used here is not uniquely defined and depends on the type of stone and mortar.

The design resistance of the masonry is determined by the stage of formation of the first cracks that cross no more than two rows [24]. In this regard, let us consider the cracking force in the mortar joint N_{crc} (**Figure 2**).

The mortar joint and the rows of bricks adjacent to it resist stretching together, provided that the necessary adhesion is provided. The crack initiation stress

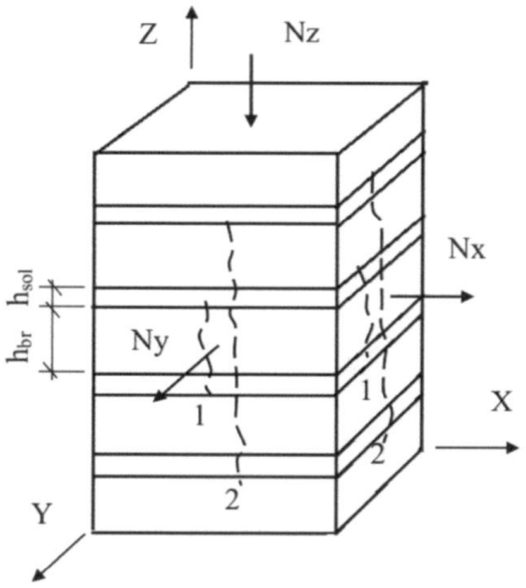

Figure 1.
Masonry element with force distribution. 1: initial stage of cracking, 2: destructive cracks.

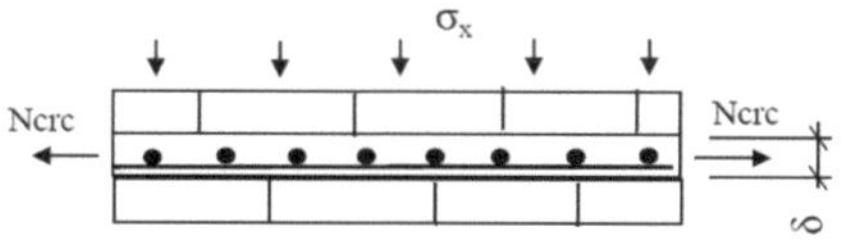

Figure 2.
Scheme of forces in the masonry layer.

corresponds to the tensile strength of the masonry over the tied joint Rt. Taking into account that usually the deformations of the mortar joint grow faster than the stone, we attribute the stress Rt only to the sum of four layers of the mortar joint, since the effect of transverse reinforcement is manifested when the grids are located at least after four rows.

The force N*crc* is resisted by the force N*s* that occurs in the reinforcement. As a result:

$$N^I = N_{crc} - N_s \tag{4}$$

Writing (4) through the mechanical parameters of materials,

$$N^I = R_t A_j - \varepsilon_s E_s A_s, \tag{5}$$

where A_j is the cross-sectional area of four mortar joints in the vertical plane; εs—deformation of the reinforcement corresponding to the deformation of the formation of cracks in the mortar joint is accepted $\varepsilon_s = \varepsilon_u$ (maximum for mortar and fine-grained concrete 1.5×10^{-4}). For composite reinforcement, the second term of the expression changes to $\varepsilon_f E_f A_f$; A_s,A_f—total cross-sectional area of reinforcement in one direction within four rows of masonry (steel and composite).

The structure of formula (6), given in SP 15.13330.2020, assumes a linear increase in the strength of unreinforced masonry R with an increase in the volumetric reinforcement coefficient μ, while a restriction is imposed $R_{sk} \leq 2R$. Simple logical reasoning leads to the fact that the strength of the reinforced masonry should increase asymptotically and not end abruptly after a linear steep takeoff.

$$R_{sk} = R + \frac{p\mu R_s}{100} \tag{6}$$

Unlike steel reinforcement used in masonry and having a physical or possibly conditional yield strength, composite reinforcement does not have such a concept, as follows from the available sources, for example, the set of rules for strengthening with composite materials SP 164.1325800.2014 "Reinforcement of reinforced concrete structures with composite materials design rules." To calculate the longitudinal reinforcement, in this case, a number of coefficients of operating conditions are introduced to the temporary resistance. Bearing in mind, the determining value for the resistance of the material of low modulus of elasticity, expression (7) can be written as follows:

$$R_{sk} = R + \frac{p\mu \varepsilon_{s,u} E_s}{100} \tag{7}$$

However, as practice shows, the stresses in steel reinforcement during the formation of cracks in the masonry, corresponding to the onset of the limit state, are still far from the design resistance of the reinforcement and its ultimate tension.

The relative deformations in (7) must be replaced by the ultimate deformations of the mortar joint in tension ε_u. Formulas (6) and (7) are comparative in nature, that is, show how much the strength of unreinforced masonry increases when it is reinforced. Therefore, this increase can be represented as a ratio of the initial bearing capacity of the mortar joint in tension to the increased bearing capacity due to the tensile resistance of the reinforcement.

In (5), the effect of reinforcement is infinite, as in the formula of the set of rules. To compensate for this shortcoming, it is proposed to introduce a restriction that would lead to an asymptote at maximum reinforcement. To do this, the decaying increase of the second term in terms of the natural logarithm function is introduced into expression (5), as the most common in analytics, and decomposed into a rapidly convergent series. Based on the general properties of the logarithm function, an argument of the form (1 + x) is introduced to exclude negative and physically nonexistent values of the function. Expression (5), taking into account the limitation on the tensile strength of the mortar joint, takes the form:

$$N^I = R_t A_j - \varepsilon_u E_s \ln(1 + A_s) \tag{8}$$

As a result, to calculate the bearing capacity of masonry reinforced with composite meshes, A.N. Plotnikov proposed a formula that takes into account the increase in the strength of unreinforced masonry due to the elastic resistance of composite reinforcement in the joints:

$$R_{sk} = R\left(\frac{R_t A_j}{R_t A_j - \varepsilon_u E_s \, ln\,(1 + A_s)}\right) \tag{9}$$

An analysis of the obtained function R_{sk} depending on A_s showed that it has an increasing and asymptotic character (**Figure 3**), starting from zero values of the cross-sectional area of the reinforcement. **Figure 3** (Graph 1) shows the dependence according to (9) of the increase in the bearing capacity of solid brick masonry on the

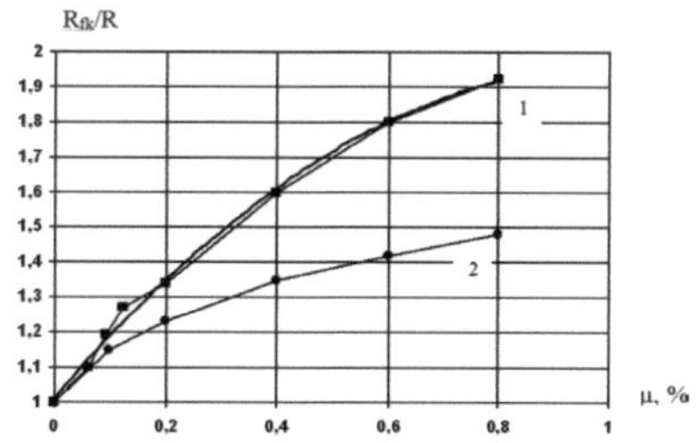

Figure 3.
Dependence of the R_{sk}/R ratio on the percentage of masonry reinforcement (1) according to the formula (8) with a logarithmic approximation, (2) according to the formula (9).

traditionally determined percentage of reinforcement. The maximum possible increase in the bearing capacity is two times.

Formula (9) can also be applied to masonry reinforced with composite rods connected into meshes, taking in the value of the elastic modulus of the composite Ef.

At the present stage of the use of composite meshes in masonry, one has to talk about a number of design limitations in determining the bearing capacity. The question of the adhesion of the composite in the body of the cement-sand mortar and, accordingly, the anchoring of the reinforcement remains unexplored. The currently used methods of connecting rods by gluing them with molten polyethylene do not give great strength. According to manufacturers, the average breaking force of the connection of rods with a diameter of 3.2 mm is Nsh = 338 N. For a masonry element with a cross section of 510 × 510 mm and a mesh of reinforcing mesh 50 × 50 mm, one rod resists shear in each of four directions from the center no more than four connections.

The modulus of elasticity of polyethylene is only about E = 300 MPa, which is significantly lower than the corresponding values of the composite rod and mortar joint. In this regard, the connections of the rods in the nodes are significantly pliable, which is reflected in the tensile strength of the reinforcement. The value of compliance can be estimated from the proportion of the location of polyethylene on the length of the rod. The length of the polyethylene section is 10 mm with a grid cell of 50 × 50 mm; that is, connection with the solution of the seam has no more than 0.8 of the length of the rod.

Compliance is also characteristic of the contact of steel reinforcement with a seam solution. The rods have the maximum compliance value at the maximum percentage of reinforcement, because at the same time, maximum stresses develop in the seams. The function of this dependence is nonlinear; in order to achieve physically defined parameters, an increasing function of the type is proposed with the introduction of compliance $k = \cos^5 x$. It has limits at x = 0: k = 1, at x = 1: k = 0.5. It is proposed to use the traditional reinforcement factor $\mu \le 1$ expressed in radians as the function argument. As a result, we get:

$$R_{fk} = R\left(\frac{R_t A_j}{R_t A_j - \cos^5 \mu \varepsilon_u E_f \ln\left(1 + A_f\right)}\right) \tag{10}$$

Figure 3 (Graph 2) shows the dependence of the increase in masonry strength depending on the percentage of reinforcement, and the maximum increase is achieved by 1.5 times.

The analytical dependence was verified by testing samples of masonry reinforced with steel and composite meshes in the joints.

The dimensions of the samples in the section are 0.51 × 0.51 m. A sample with steel meshes was a prism with a height h = 1.34 m. Ceramic bricks of the M125 brand were used on a cement-sand mortar of the M100 brand; reinforcement was made with meshes of wire Ø4 Vr500 with a cell measuring 50 × 50 mm, laid horizontally every three rows of bricks (**Figure 4**).

Material parameters: ultimate strength of brick in bending R_{ben} = 2.6 MPa, in compression R = 13.4 MPa; cement-sand mortar grade M100 with cubic strength R = 10 MPa; reinforcing wire Ø4 Vr500 (As = 12.57 mm^2) normative tensile strength R_{sn} = 500 MPa, calculated—R_s = 415 MPa. The masonry was created immediately on the press plate.

To determine the physical and mechanical characteristics of the working reinforcement, tensile tests were carried out. For the purpose of carrying out subsequent measurements, calibration dependence was built for strain gauges.

To measure deformations and stresses in the rods as part of the structure, strain gauges with a base of 20 mm and a resistance of 100 Om were glued to them. The strain gauge glued to the rod was covered with sealant, and the wires were removed from the masonry. The sensors were located on two grids: above the ninth and fifteenth rows of masonry (**Figure 5**).

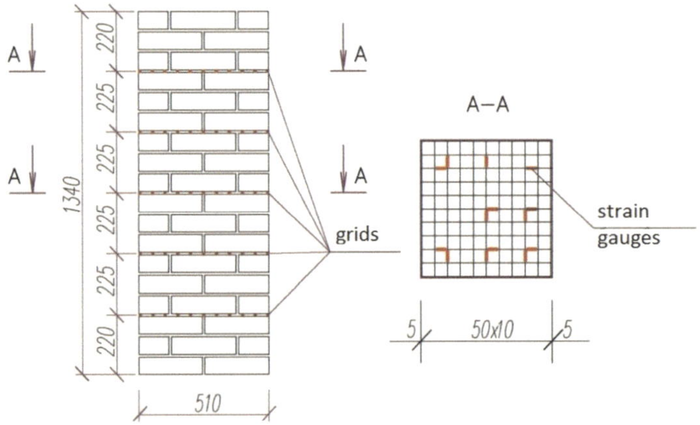

Figure 4.
Arrangement of reinforcing meshes and sensors.

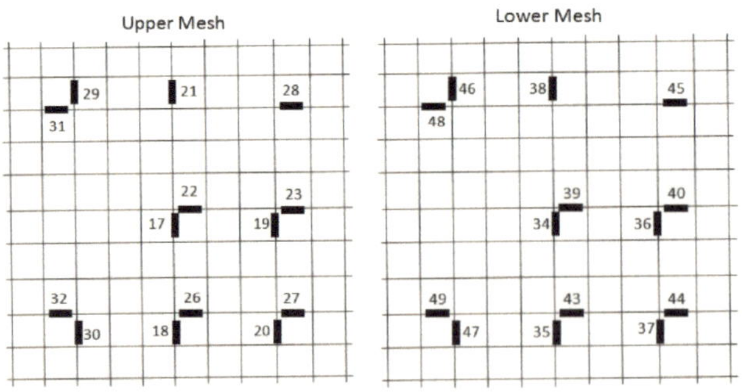

Figure 5.
Location of strain gauges on reinforcing masonry meshes.

During the test, the following measuring instruments were used:

- deflection meters Aistov 6-PAO (P1–P4) with an accuracy of 0.01 mm (**Figure 6**) to assess the total vertical deformations of the masonry column;
- mechanical linear meters (M1–M4), with an accuracy of 0.01 mm for measuring longitudinal deformations of the masonry;
- electronic strain gauges DPL-10 with connection to the recorder “Terem-4.0” with an accuracy of 0.001 mm for measuring surface deformations of the masonry:

  a. D1, D4, D6, D9—for measuring the longitudinal deformations of the masonry (duplicating M1–M4);

  b. D2, D3, D5, D7, D8, D10—for measuring the transverse deformations of the masonry.

On the general views of the sample (**Figure 7**), the numbers of mechanical deflection meters for measuring vertical deformations, electrical strain gauges for transverse and longitudinal deformations of the masonry are indicated. AID-5 recording equipment was used.

The dimensions of the cross section of the masonry (510 mm) were sufficient to determine the deformations along the width of its section.

Loading was carried out in steps of 200 kN with central compression on a hydraulic press with a capacity of 5000 kN. At each stage, the load was kept for at least 10 minutes. Longitudinal strains were measured using mechanical gauges mounted on a base 455 mm high on all four sides; longitudinal and transverse deformations by electrotensometers with a base of 150 mm.

Comparison of the work of solid brick masonry reinforced with composite mesh with reinforcement with traditional steel mesh (Vr500 wire) was carried out on samples with dimensions of 380 × 380 × 600 mm with the same percentage of reinforcement.

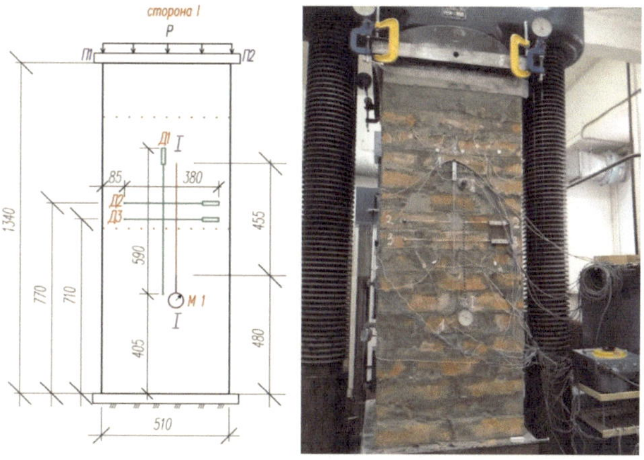

a) b)

Figure 6.
Test stand. a—general scheme; b—general view.

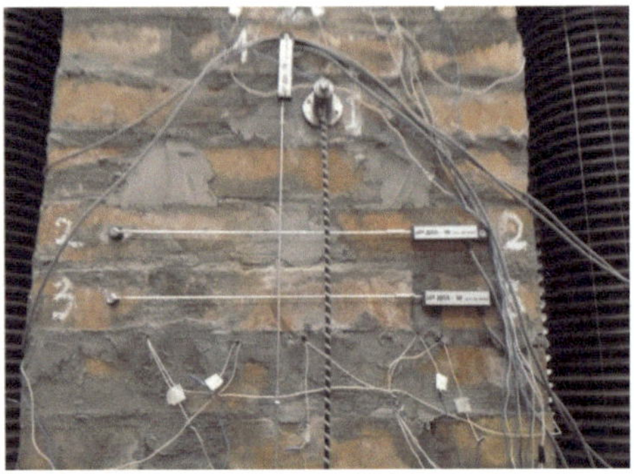

Figure 7.
Placement of strain gauges on the sample surface.

3. Results and problems

The greatest interest in the tests carried out was the distribution of stresses in the reinforcing bars of the meshes over the cross section of the masonry. According to preliminary calibration graphs and data measured during loading from strain gauges on reinforcing meshes, the forces and stresses in the rods of the masonry mesh were determined. The stresses in the central and peripheral parts of the reinforcing mesh depending on the load are shown in **Figures 8** and **9**.

According to the test results, it was determined that the stresses in the reinforcing bars are 37% in the center of the masonry and 20% in the peripheral sections of the designed steel resistance. Stresses along the height of the sample are distributed unevenly. In the upper grids, the stresses in the center of the masonry section are 1.36

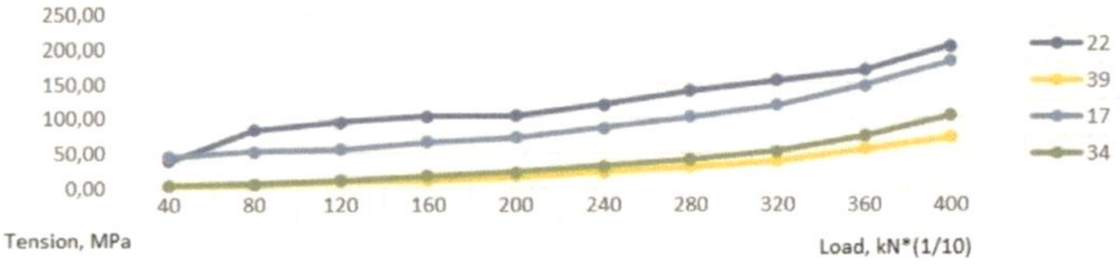

Figure 8.
Stresses in the central part of reinforcing meshes.

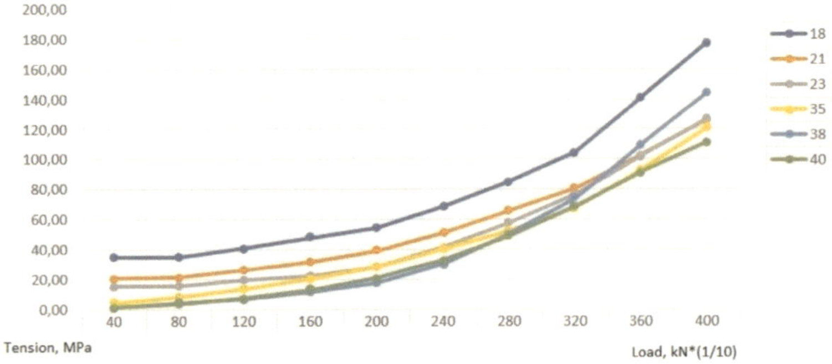

Figure 9.
Stresses in the peripheral part of reinforcing meshes.

times higher than the values of the lower grid. In the peripheral zones of the upper grids, the voltage is 1.33 times higher.

During the formation of cracks in the masonry, the maximum stresses in the rods were 92 MPa at the center of the section and 48 MPa at peripheral points at the corners of the masonry.

Up to a load of 2000 kN, the stresses in the rods increase linearly, above the stresses increase nonlinearly, while cracks in the masonry are not yet formed. This indicates the plastic nature of the work of the masonry; that is, there is a collapse of the mortar joint under the action of reinforcing bars and there is some movement of the stones relative to the mortar joints.

There is a margin of bearing capacity for tension of reinforcing bars. The norms specify the resistance of the bars as 0.6 Rsn. It is determined, according to the test, that this value is higher and is 0.72 Rsn.

Longitudinal strains were measured on four sides of the sample (**Figure 10**).

Longitudinal deformations are determined equally for all groups of sensors (mechanical and electronic).

The transverse deformations of the masonry in the reinforced and non-reinforced layers have a nonlinear nature of work, which indicates an increase in the intensity of cracks inside the volume of the masonry (**Figure 11**).

Theoretical values of the bearing capacity of unreinforced and reinforced samples of the considered sizes were $N_{ur} = 624$ kN,. and $N_u = 973$ kN respectively, an increase of 1.56 times. Strength limit of reinforced masonry $N_u = 1600$ kN

The load at which the destruction of the sample began was 4020 kN. The margin of bearing capacity is 2.5 times. This reserve can be attributed to a different technology for the manufacture of masonry in comparison with the stipulated norms and created in the laboratory. The design resistance of the reinforced masonry according to the formula (10) R_{sk} = 14.85 MPa. Four rows of masonry of the experimental sample are taken into account. Seams are accepted with a thickness of 1 cm. Eleven wire rods are located in one horizontal seam.

Numerical data: ε_u=1.5*10^{-4}—reinforcement deformation corresponding to the deformation of mortar joint crack formation (ultimate deformations); A_j —cross-sectional area of four mortar joints in the vertical plane, A_j = 51*1*4 = 204 cm^2; A_s is the total cross-sectional area of reinforcement in one direction within four rows of masonry, $A_s = 12{,}57 * 10^{-2} * 11 = 1{,}3827$ cm^2; R_t is the tensile strength of the masonry

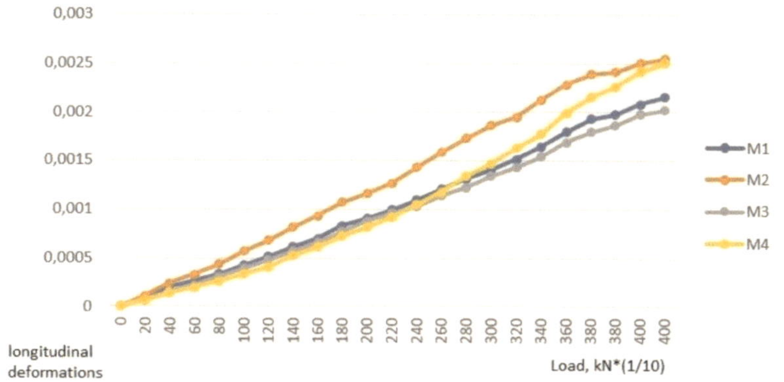

Figure 10.
Longitudinal deformations of masonry on four sides of the sample.

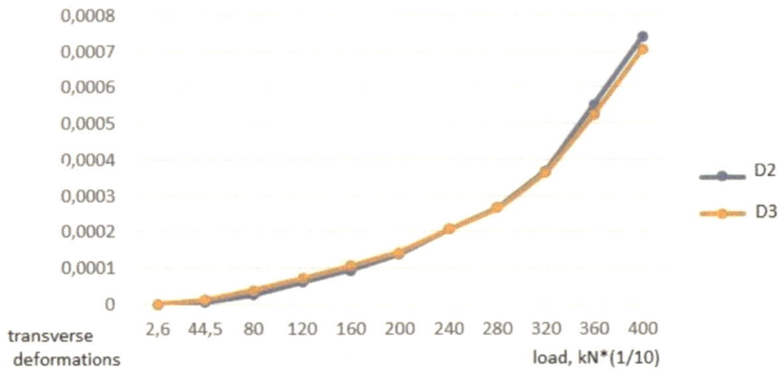

Figure 11.
Transverse masonry deformations. D3—closer to the reinforced layer, D2—to the layer without reinforcement.

along the tied seam, R_t = 0.16 MPa (according to SP 15.13330.2020); E_s = 200,000 MPa—elastic modulus of reinforcing steel.

In this case, N_u = 3862 kN, which is close to the ultimate test load of 4020 kN.

The relationship between stresses and strains was nonlinear. The initial deformation modulus of masonry with mesh reinforcement according to SP 15.13330.2020 (6.21) is taken to be the same as for unreinforced:

$$E_0 = \alpha R_u, \tag{11}$$

According to the results of the experiment, for masonry of solid ceramic bricks at α = 1000, E_0 = 15,450 MPa is taken. According to the results of measurements, E_0 = 10,666 MPa.

Poisson's ratio for the area of deformations in the first third of the increase in load

$$\nu = \frac{\varepsilon_x}{\varepsilon_y}, \tag{12}$$

where ε_x and ε_y are the relative transverse and longitudinal strains, respectively.

At a maximum load of 4020 kN, the deformation modulus E = 6925 MPa. A decrease in E as a result of the nonlinearity of the processes was noted by 1.6 times.

The increase in Poisson's ratio for reinforced masonry was insignificant and amounted to 0.23, compared with the standard ν = 0.25, by less than 10%.

The magnitude of the absolute vertical deformation of the sample was Δ_y = 4.17–4.58 mm. Relative vertical deformations $\varepsilon_y = 3.11*10^{-3} - 3.41*10^{-3}$.

The effect inherent in the reinforcement of the masonry with meshes in the mortar joints in the test proved to be quite complete. The pattern of masonry cracking changes, and no main cracks appear. Visible cracks occur in the brick in the layer above the mesh. A material with a high modulus of elasticity increases the resistance of mortar joints and masonry in general.

In the previous experiments [24, 25], at maximum loads, small fragments of brick and mortar peel off, which does not occur in unreinforced masonry. At the same time, stresses in the mesh rods reach the yield strength of steel before the failure of the masonry.

In masonry samples using composite meshes, this phenomenon is not observed or it is less pronounced. This is explained by the significantly lower value of the elastic modulus of the composite relative to steel.

Comparison of the work of solid brick masonry reinforced with composite mesh with reinforcement with traditional steel mesh (Vr500 wire) was carried out on samples with dimensions of 380 × 380 × 600 mm with the same percentage of reinforcement.

Comparison of numerical data obtained by formulas (9) and (10) was carried out with a number of experimental data. The test results are mentioned in the following studies: V.M. Pozdeev, N.P. Soloviev, A.V. Vinogradov, V.V. Nikolaev [22]; A.B. Antakov [23]; A.V. Granovsky, V.V. Galishnikova, E.I. Berestenko [21] (**Figure 12**).

The most relevant information on this experiment was obtained from a comparison of the transverse deformations of the masonry (**Figure 13**), determined near the mortar joint.

It has been found that the transverse deformations of masonry pillars reinforced with composite, in particular, basalt-plastic reinforcement (FRP) are 2.5 times higher

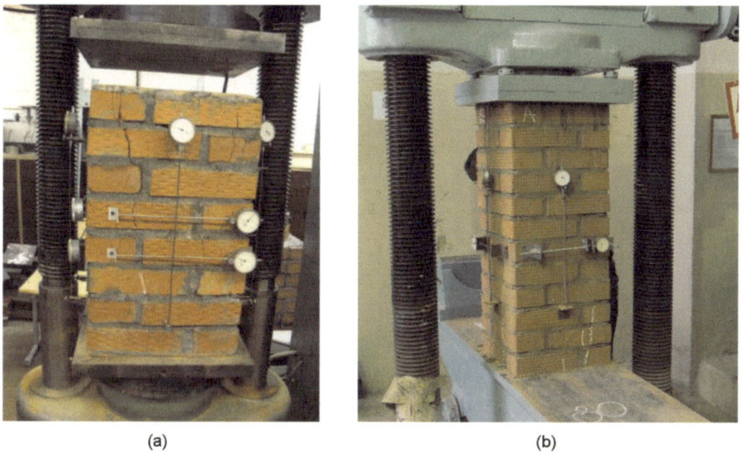

Figure 12.
Prototypes with measures placed on them: (a) tests by V.M. Pozdeev and (b) tests of A.B. Antakov.

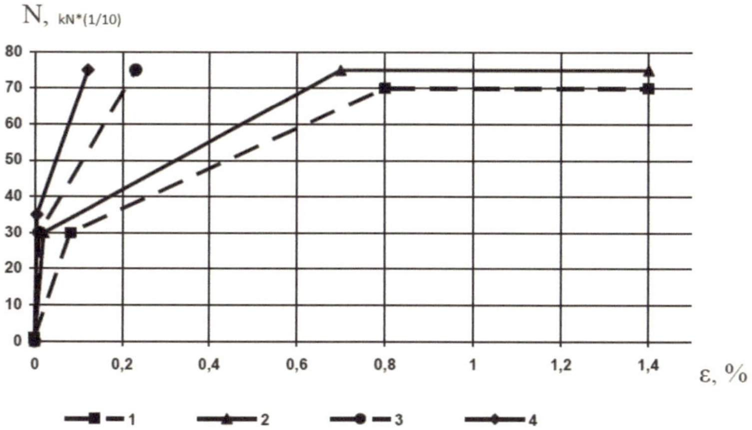

Figure 13.
Graph of transverse deformations of columns and reinforcement in the central cross section according to the tests of V.M. Pozdeev: (1) transverse deformations of columns with FRP; (2) transverse deformations of columns with steel reinforcement; (3) FRP elongation; and (4) elongation of steel reinforcement.

than steel ones. The tensile strength and cracking load are practically the same. Such results were obtained with a relatively small percentage of reinforcement –0.11% and a solution that did not gain full strength, with early cracking.

In relation to unreinforced masonry, reinforcement with composite reinforcement with different percentages of reinforcement according to the test results [23] in samples 380 × 380 × 1000 mm led to an increase in the bearing capacity and crack resistance by 30–33%. The intensity of reinforcement varied in the range of 0.062–0.422% (**Figure 14b**). In all series, the destruction of masonry with composite

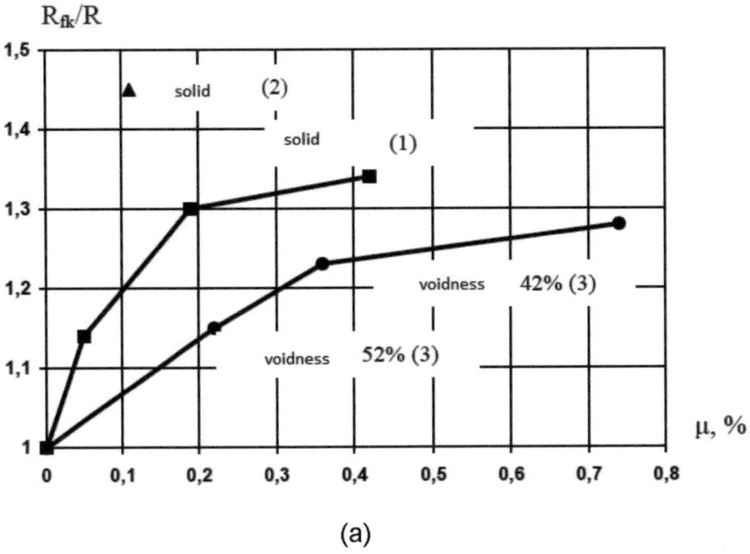

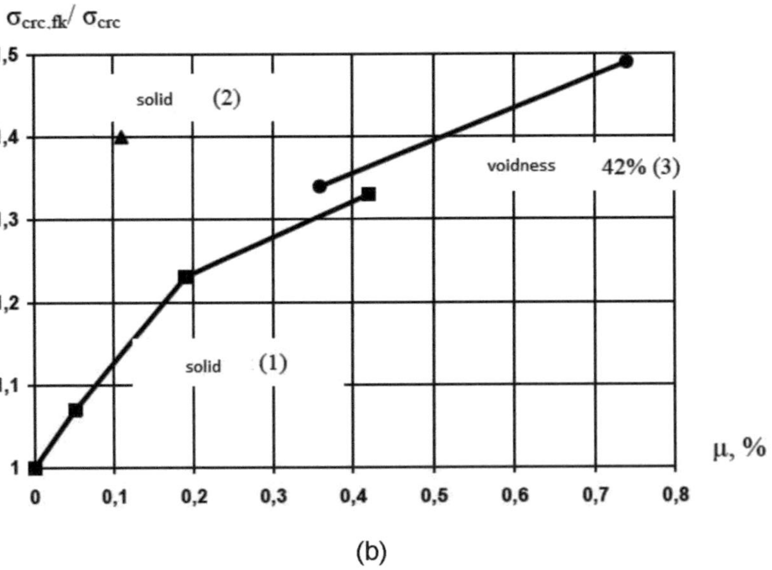

Figure 14.
Ratio dependence: (a) R_{fk}/R;)b) $\sigma_{crc,fk}/\sigma_{crc}$ of the percentage of reinforcement of masonry composite according to experimental data: (1) A.B. Antakov, (2) V.M. Pozdeev, (3) A.V. Granovsky.

reinforcement took place with a main vertical crack, which is not typical for reinforcement with steel meshes [24].

Tests conducted under the leadership of A.V. Granovsky [21], reinforced with composite reinforcement masonry made of ceramic stones with large voids and small sections (sample sizes 250 × 1030 × 1200 mm, 250 × 800 × 1350 mm), showed an increase in the bearing capacity relative to unreinforced ones by 1.2–1.3 times (**Figure 14a,b**).

All experimental data show lower strength values of composite-reinforced masonry compared with steel-reinforced masonry. The reason for this is the lower modulus of elasticity of the plastic, which does not help to contain the transverse deformations of the masonry. However, this disadvantage can be overcome by increasing the percentage of reinforcement by the composite. Another reason is the insufficient adhesion of the surface of plastic rods with the joint solution and the ductility of the nodal joints of composite meshes made in particular with polyethylene. In the case of using more rigid connecting materials, the bearing capacity of the masonry reinforced with a composite can be increased by 1.3 times, as follows from the graph (**Figure 13**). To increase the bearing capacity of such masonry, a thinner mortar joint with the same percentage of reinforcement as with steel meshes can be achieved due to small diameters and frequent placement of rods.

4. Conclusion

To determine the bearing capacity of masonry, primarily reinforced with steel and composite meshes in horizontal joints, it is necessary to use the characteristics of the source materials, using the obtained analytical dependence. It is recommended to use not the design resistance of the reinforcement material, but its modulus of elasticity and the value of ultimate deformation. It gives a convergence with experiments of 4%.

According to the test results of masonry with steel mesh, in comparison with the current standards, a 2.5 times greater strength was obtained. For composite reinforcement, there is no information in the norms.

Compared with unreinforced masonry, steel mesh reinforcement increases strength by a maximum of two times. According to the results of generalized tests, reinforcement with composite meshes increases the bearing capacity of masonry, depending on the types of composites used, by 1.3–1.5 times. From samples with steel reinforcement, this is 65—75%.

No main cracks were formed in the sample of masonry with steel reinforcement. The destruction occurred along small chips of brick and mortar. Stresses in the reinforcing bars of steel meshes did not reach the yield strength of steel and amounted to 37% in the center of the masonry and 20% along the perimeter. During the formation of cracks, they amounted to 92 MPa and 48 MPa, respectively.

The deformation modulus of the reinforced masonry during loading decreased by 1.6 times.

An increase in the bearing capacity of masonry reinforced with composite meshes is possible due to structural improvements, primarily by connecting rods at intersections.

Author details

Alexey N. Plotnikov[1*], Viktor A. Ivanov[1], Boris V. Mikhailov[1], Tatyana G. Rytova[2], Olga S. Yakovleva[1], Mikhail Yu Ivanov[1] and Natalia V. Ivanova[1]

1 Chuvash State University, Russia

2 Moscow State University of Civil Engineering, Russia

*Address all correspondence to: plotnikovan2010@yandex.ru

References

[1] Ivanov VA. Numerical modeling of the dynamics of structures supported by a system of reinforcing rods. In: Ivanov VA, Vutsin LI, Skobeeva MV, editors. Modern Problems of Continuum Mechanics 2019: Collection of Articles. Art. Basedonthe Materials of the Conference from the International Participation. Cheboksary: Chuvash Publishing House University; 2019. pp. 119-124

[2] Ivanov VA, Kibets AI, Kibets Y. Finite element technique for solving a three-dimensional problem of the dynamics of structures reinforced by a system of reinforcing elements. Problems of Strength and Plasticity. 2019;**81**(2):191-201

[3] Babaeidarabad S. Masonry Walls Strengthened with Fabric-Reinforced Cementitious Matrix Composite Subjected to In-Plane and Out-of-Plane Load. 2013

[4] Akhaveissy AH. The DSC model for the nonlinear analysis of In-plane loaded masonry structures. The Open Civil Engineering Journal. 2012;**6**:200-214

[5] Nwofor TC. Experimental determination of the mechanical properties of clay brick masonry. Canadian Journal on Environmental, Construction and Civil Engineering. 2012;**3**(3):127-145

[6] Antakov AB, Sokolov BS. Analytical assessment of the stress-strain state of masonry under compression based on the author's theory. Building Materials. 2019;**9**:51-55

[7] Sokolov BS, Antakov AB. Research results of stone and reinforced stone masonry. Bulletin of National Research Moscow State University of Civil Engineering. 2014;**3**:99-106

[8] Kashevarova GG. A model of a masonry wall for the study of schemes and mechanisms of destruction. In: Kashevarova GG, Vildeman VE, Akulova AN, editors. Information, Innovations, Investments: Collection of Articles. Materials Conference. Perm. 2002. pp. 38-41

[9] Pangaev VV. Development of computational and experimental methods for studying the strength of masonry of stone structures. In: Pangaev VV, editor. Abstract for the Application: Art. Doctors of those Sciences. Novosibirsk; 2009

[10] Kabantsev OV. Scientific Foundations of the Structural Theory of Masonry for Assessing the Limiting States of Stone Structures of Earthquake-Resistant Buildings. Moscow; 2016

[11] Sokolov BS. Development of methods for calculating stone and reinforced stone structures. In: Sokolov BS, Antakov AB, editors. New in Architecture, Design of Building Structures and Reconstruction: Materials of the IV International (X All-Russian) Conference NASKR-2018. Cheboksary: Publishing House of Chuvash University; 2018. pp. 174-183

[12] Kapustin SA, Likhacheva SYu. Modeling the Processes of Deformation and Destruction of Materials with a Periodically Repeating Structure. Nizhny Novgorod: NNGASU publishing house. 2012. p. 96

[13] Ali SS. Finite element model for masonry subjected to concentrated loads. Proceedings of the American Society of Civil Engineering: Journal Structural Division. 1990;**114**: 1761-1784

[14] Antakov AB, Plotnikov AN, Pozdeev VM. Bearing capacity of masonry reinforced with grids made of basalt-plastic reinforcement. In: Tamrazyan AG, Kopanitsa DG, editors. Modern Problems of Calculating Reinforced Concrete Structures, Buildings and Structures for Emergency Impacts. Moscow: National Research Moscow State University of Civil Engineering; 2016. pp. 15-21

[15] Kaushik HB. Uniaxial compressive stress-strain model for clay brick masonry. Current Science. 2007;**92**: 497-501

[16] Plotnikov AN. Strength calculation of reinforced masonry based on deformation parameters of its constituent materials. In: Plotnikov AN, Yakovleva OS, Romanova TV, editors. Modern Problems of Continuum Mechanics—2019: Collection of Articles. Art. Based on the Materials of the Conference from the International Participation. Cheboksary: Publishing House "Wednesday"; 2019. pp. 60-68

[17] Plotnikov AN. Bearing capacity of reinforced masonry under central compression based on the deformation parameters of its components. In: Plotnikov AN, Romanova TV, Mikhailov BV, Yakovleva OS, Yu M, editors. Construction and Development: Life Cycle – 2020: Materials of the V International. Cheboksary; 2020. pp. 183-197

[18] Plotnikov AN, Romanova TV, Mikhailov BV, Yakovleva OS, Ivanov MY. Rearing capacity of reinforced masonry under central compression based on the deformation parameters of its components. In: Vatin NI, Tamrazyan AG, Plotnikov AN, Leonovich SN, Pakrastins L, Rakhmonzoda A, editors. Advances in Construction and Development. Singapore: Springer; 2022, 2020

[19] Stepanova VF, Buchkin AV, Yu E. Composite polymer mesh for masonry. Building Materials. 2019;**9**:44-50

[20] Stepanova VF, Buchkin AV, Ishchuk MK, Granovsky AV. Composite polymer mesh for masonry. Industrial and Civil Construction. 2019;**11**:15-19

[21] Granovsky AV, Galishnikova VV, Berestenko EI. Prospects for the use of reinforcing mesh based on basalt fiber in construction. Industrial and Civil Construction. 2015;**3**:59-63

[22] Vinogradov AV. Study of the possibility of using basalt-plastic reinforcement for reinforcing masonry, In Interuniversity: Sat: Innovative Resources and National Security in the Era of Global Transformations, Yoshkar-Ola: MarGTU, 2012, 144–146

[23] Antakov AB. The strength of masonry reinforced with composite meshes. Successes of Modern Natural Science. 2014;**7**:116-120

[24] Onishchik LI. Strength and Stability of Stone Structures. Part 1. Moscow. 1937. p. 291

[25] Onishchik LI. Stone Structures for Industrial and Civil Buildings. Moscow. 1939. p. 208

Chapter 6

Experimental Investigation on Clay Bricks Using Babul Sawdust Bricks

Praveen Kumar R., Balaji D.S. and Navaneethakrishnan G.

Abstract

The construction practices of today demands production of alternative building materials, which consume less energy and can be used for construction. One such material is the babul tree sawdust bricks. In this work, the babul sawdust is prepared using the locally available babul tree in India. Hence, an attempt is made to stabilize these blocks using clay and sawdust. The saw dust percentage has been varied from 0 to 50% by weight. The results show the variation in properties such as compressive strength, initial rate of absorption and water absorption are studied and compared.

Keywords: babul sawdust, clay, compressive strength, water absorption, sawdust bricks

1. Introduction

Earth has been the most widely known and abundantly available material for human society to use it in construction. From the days of Egyptian and Mesopotamian earth is main part of any construction in its different forms [1]. Nowadays, several research fields on materials recycling environmentally friendly and energy conservation are operated. Many previous researches have obtained valuable results to use the industrial wastes in various forms of construction materials production [2]. So we are used babul sawdust in manufacturing of bricks. In addition, demand for clay bricks with higher insulating capacity is increasing. For this purpose, we used babul sawdust and other organic materials most frequently used as pore formers [3]. These materials had properties which resembled those of lightweight brick materials. The Babul sawdust is the byproduct of sawing babul tree timbers. The recycling of the wood chips such as sawdust which offers the required properties of ceramic products. The chemical composition of the sawdust is 60.8% of carbon, 33.83% of oxygen, 5.19% of Oxygen and 0.90% of Nitrogen [4]. In this study, investigation of the sawdust suitability to use in combination of ceramic material was carried out. The clay bricks made with the mixture of sawdust and ceramic material have advantage compared to traditional bricks in the aspect of action of degreasing, low density and alveolar appearance, improved mechanical strength. Various experimental works and reviews related to the study of saw dust have been carried out [5].

The cohesive nature of the clay imparts plasticity to the soil under moist conditions. The thin film of water absorbed ensures the strong adherence between the layers leads to plasticity. The mineral present in the clay acts as a natural binding agent.

Figure 1.
Babul tree and saw dust.

The affinity of the clay towards water results in swelling and shrinking when it dries, especially it is prominent when montmorillonite is present. Stability agents like lime added to the soils with the clay content of above 30%. Particle size is ranges from less 0.002 mm to greater than 2 mm. Babul tree known for the exploitation of the ground water and its impact on reduction of the water table. Even it grows in the drought hit areas with no ground water by absorbing the water molecules in the air (humidity), leaving the place dry and affects the rainfall also. The roots of the babul tree destroy the soil nutrients. It produces carbon dioxide more than the oxygen generation which makes it unlikely even for the birds to have their shelter. The seeds and the parts of the babul tree is of no use to the humans and animals. Earlier the babul tree seed was sowed in various drought hit regions of India for firewood purpose. After knowing the ill effects on the environment, many global organizations steps forward to create awareness. The Babul Tree and saw dust is shown in **Figure 1**.

2. Experimental work

2.1 Specific gravity test of sawdust

It is defined as the ratio of the density of any substance to the density of some other substances taken as standard, water being the standard for liquids and solids, and hydrogen or air being the standard for gases. Weigh a clean and dry le chatelier flask of bottle with its stopper noted as W1. Clay sample filled half of the flask (about 50 gram) and weigh it with its stopper noted as W2. Add water to in the flask till it is half full. Mix with glass rod thoroughly to remove entrapped air. Continue stirring and add more water till the graduated mark. The Specific gravity test instrument is shown **Figure 2**. Then the pychnometer is completely filled with water, wiped of the outside and weighed again W3. The pychnometer is then emptied and filled with water and weighed W4.

$$\text{Specific gravity} = \frac{(w_2 - w_1)}{(w_2 - w_1) - (w_3 - w_4)} \times 100 \quad (1)$$

Weight of empty bottle, $w_1 = 0.673$
Weight of soil, $w_2 = 1.22$
Weight of soil and water, $w_3 = 1.83$
Weight of water, $w_4 = 1.5$

Figure 2.
Specific gravity test instrument.

2.2 Sieve analysis test of sawdust

A sieve analysis is a procedure used to assess the particle size distribution of a granular soil. It is performed on any type of non-organic or organic granular material including sands, crusher rocks, clays, granite, feldspars, soil, coal, grain and seeds down to a minimum size depending on the exact method. About 1000 grams of oven dried soil retained as 75 micron sieve is taken. The soil is sieve through the set of sieves as per the order of arrangement indicated sieve sizes: 4.75 mm, 2.36 mm, 1.18 mm, 600 μ, 425 μ, 300 μ,150 μ,75 μ and pan. The cover is placed over the top of stack of the sieves. The set of sieves is shaken for about 10 minutes giving both horizontal and vertical movements. The soil retained in each sieve is transferred to separate plates and weighted accurately. Cumulative weight retained cumulative percentage retained and percentage passing are calculated. The Sieve analysis instrument is shown in **Figure 3**.

$$\text{Percentage of retained} = \frac{\text{weight of material retained in each seive}}{\text{weight of sample taken for the test}} \times 100 \quad (2)$$

$$\text{Percentage of passing} = 100 - \text{Percentage of retained}$$

$$\text{Effective size of clay} = 90\,\text{microns}$$

2.3 Liquid limit test of sawdust

The liquid is arbitrarily defined as the water, in percent at which a part of soil in a standard cup and cut by a groove of standard dimensions will flow together. Weighed about 120 g of soil passing through 420 μ I.S sieve. The soil sample is placed on the evaporating dish and thoroughly mixed with water using spatula. The casagranda's device is checked to have a correct fall of 10 mm and placed a portion of the prepared paste over the brass cap. The groove is made in the middle of

Figure 3.
Sieve analysis instrument.

the soil cake using the grooving tool. It is rotated at the rate of 2 blows per second and the relations are counted until the groove closes over a length of 12 mm. At center of test sample, a small quantity is collected in a container and its weight is noted. The sample is dried in the oven for 24 hrs. and weighed. The difference of the two weights will give the moisture content. The experiment is repeated by adding more water. Four trials are made, so that the numbers of blows are more than 25 in two cases and less than 25 in other two cases. In each trial moisture are determined. The Liquid limit test results shown in **Table 1**. The Liquid limit test instrument is shown in **Figure 4**.

2.4 Plastic limit test of sawdust

The plastic material is defined as the moisture content at which the plastic material can be molded into a shape and the material will retain that shape. If the moisture content is below the plastic limit, it is considered to behave as a solid material. A sample of about 50 gram is taken in a glass plate and mixed thoroughly with water, rolled into ball shape and made into thread with a diameter of 3 mm. The process of making thread by kneading and rolling again is repeated until the soil ceases to be

Weight of dry soil (gms)	Quantity of water	Percentage of water added	Number of blows
120	22	18	112
120	26	21	73
120	30	25	55
120	32	26	26
120	34	28	13

Table 1.
Liquid limit test results.

Figure 4.
Liquid limit test instrument for measurement.

Observation (gm)	Trial 1	Trial 2
Weight of can (W_0)	11.7	13.4
Weight of wet soil with can (W_1)	12.6	14.6
Weight of dry soil with can (W_2)	12.4	14.3
Weight of water (W_2–W_1)	0.2	0.3
Weight of dry soil (W_2–W_0)	0.7	0.9
Moisture content (W%)	28.57	33.3

Table 2.
Tabulation for plastic limit test.

plastic and crumbles. The sample of the crumbled soil was collected together and placed in a container. The test is repeated twice more with fresh samples. The average of the three water contents gave the plastic limit value. The Tabulation for plastic limit test **Table 2**. The Plastic limit test instrument as shown in **Figure 5**.

2.5 Shrinkage limit of sawdust

The shrinkage limit is the maximum water content at which a reduction in water content does not significantly reduce the volume of the soil mass. After a certain point, when the water content continues to drop, air begins to seep into the soil's voids, maintaining the void's volume. Mix 30gm of soil that has passed through a 425 μ sieve with distilled water.. Without adding air bubbles, the water should be enough to make the soil pasty in the shrinkage dish. As soon as the shrinkage dish is filled with red soil, weigh it. The dish should be dried in both the air and an oven. With the dry soil paste, weigh the shrinkage dish. Determine the dish's empty mass after cleaning and drying it. Weigh a second, empty, ceramic dish that will be used to measure the weight of mercury. Keep the shrinkage dish inside a sizable porcelain dish, overflow it with mercury, and scoop out the extra by pressing a plate of plain glass firmly over the dish's top. Wipe the outside of the glass cup to remove any adhering mercury, and then place it in another dish. Place a dry soil paste on the surface of the mercury and submerge it under the mercury by pressing with glass plate with

Figure 5.
Plastic limit test instrument.

Trial no	1	2	3	4
Water content (%)	65.77	79.55	63.94	69.75
Shrinkage limit (%)	12.11	9.04	9.26	9.18

Table 3.
Tabulation for shrinkage limit.

prongs. Transfer the mercury displaced by the soil paste to the mercury weighing dish and weight. The tabulation for Shrinkage limit as shown in **Table 3**. The Shrinkage Limit Instrument is shown in **Figure 6**.

2.6 Compressive strength test of bricks

This test is carried out to determine the brick's compressive strength. It is also known as the brick's crushing strength. Six brick samples are typically brought to a laboratory for testing and examined one by one. A brick specimen is placed on a crushing machine during this test, and pressure is applied until the brick breaks. It is taken into account the maximum pressure at which bricks are crushed. Each of the six specimens is tested separately, and the average result is used to determine the compressive strength of bricks. Make a note of the specimen's dimensions. The specimen should be placed between compression grips. Apply the load now. Gradually raise the load and record the load at which the specimen fails. Divide the load by the contact surface area to determine the compressive strength. Find the materials' average compressive strength by testing three specimens. **Table 4** displays the outcomes of the compression strength test. **Figure 7** depicts the Brick in a loading condition. **Figure 8** displays the Compressive Strength Chart.

Figure 6.
Shrinkage limit instrument.

Saw dust	Compressive strength value			Average (N/mm^2)
	S1	S2	S3	
0%	7.1	7.3	7.35	7.25
10%	7.55	7.9	7.8	7.75
20%	7.6	7.8	7.46	7.62
30%	7.27	7.55	7.26	7.36
40%	7.27	7.01	7.2	7.16
50%	6.95	7.01	7.04	7

Table 4.
Compressive strength test results.

2.7 Water absorption test of bricks

In this test, dry bricks that have been weighed are submerged in fresh water for 24 hours. Following immersion, the items are removed from the water and dried with a cloth before the brick is weighed while still wet. The water detected by brick accounts for the weight discrepancy. Next, the water absorption is computed. Brick's quality increases with how little water it absorbs. An excellent brick will not absorb 20% of its own weight. The sawdust bricks used in **Figure 9** water absorption tests. The results of the water absorption test are displayed in **Table 5**.

2.8 Efflorescence test of bricks

Alkalies in bricks are harmful, and by absorbing moisture, they turn the surface of bricks gray or white. This test is carried out to determine whether alkalies are

Figure 7.
Brick under loading condition.

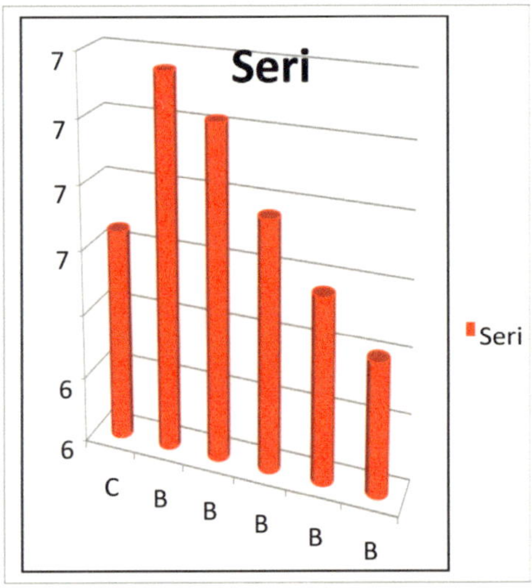

Figure 8.
Compressive strength chart.

present in bricks. In this experiment, a brick is submerged in fresh water for 24 hours, removed, and then given time to dry into the desired shape. It is evidence that there are no alkalies in brick if the whitish layer is not visible on the surface. The presence of alkalies is acceptable if it covers about 10% of the brick surface and is visible. It is moderate if that represents 50% of the surface. Alkalies have a significant negative impact on brick if they are present in excess of 50%. The Brick after Efflorescence test as shown in **Figure 10**. The results of Efflorescence test as shown in **Table 6**.

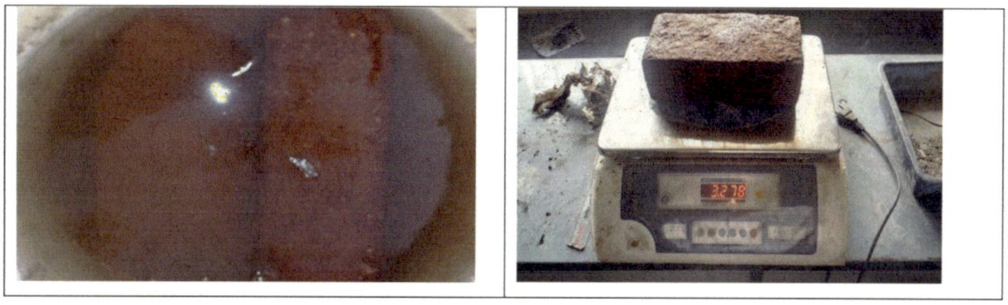

Figure 9.
Sawdust bricks during water absorption test.

Sawdust percentage	Weight of brick (kg)	Water absorption value (percentage)
0	3	21
10	2.91	20.1
20	2.85	18.9
30	2.8	17
40	2.77	15.5
50	2.7	14

Table 5.
Water absorption test result.

Figure 10.
Brick after efflorescence test.

2.9 Density test of bricks

All bricks' weights are measured individually. Then calculate the chamber brick and babul sawdust brick's length, width, and depth. Finally, use the following formula to determine the density of bricks. The Chamber Brick Density Test is displayed in **Table 7**. The Sawdust Brick Density Test is presented in **Table 8**.

Sawdust percentage	weight of brick(kg)	Efflorescence (gms)
0	3	0.45
10	2.95	0.45
20	2.89	0.39
30	2.83	0.33
40	2.77	0.32
50	2.7	0.35

Table 6.
Tabulation of efflorescence test.

S. No	Sample	Size (m)			Weight (kg)	Volume (m^3)	Density (kg/m^3)
		L	B	D			
1	1	0.22	0.10	0.08	3	0.0017	1764.70
2	2	0.22	0.10	0.08	3.15	0.0017	1789.77
3	3	0.22	0.10	0.08	3.05	0.0017	1794.11

Table 7.
Density test for chamber brick.

S.No.	Sawdust (%)	Size(m)			Weight (kg)	Volume (m^3)	Density (kg/m^3)
		L	B	D			
1	0	0.22	0.10	0.08	3	0.0017	1764.70
2	10	0.22	0.10	0.08	2.91	0.0017	1711.76
3	20	0.22	0.10	0.08	2.85	0.0017	1676.47
4	30	0.22	0.10	0.08	2.8	0.0017	1647.05
5	40	0.22	0.10	0.08	2.77	0.0017	1629.42
6	50	0.22	0.10	0.08	2.7	0.0017	1588.23

Table 8.
Density test for sawdust bricks.

3. Conclusion

- Effective utilization of the available resources without harming the environment is a major concern in the present situation.
- Many countries takes various initiatives through implementation of policies and awareness programs.
- Considering the ill effects of the babul tree on the environment, an alternate method which is environmentally friendly to utilize its byproducts (sawdust) is important. Based on the investigation, the Babul sawdust seems to be a potential replacement for the clay in brick making.

- The findings show that about 50% of babul sawdust is the ideal amount to replace clay soil, producing a block with a high compressive strength value of 7 N/mm^2 with water absorption value of 14% and efflorescence value of 0.35. Compared to regular bricks, this brick has a high efficiency and cost-effective. We are using this tree sawdust as a replacement material of clay in brick.

- Babul sawdust may one day be able to partially replace clay in the production of brick blocks.

Author details

Praveen Kumar R.[1*], Balaji D.S.[2] and Navaneethakrishnan G.[3]

1 Department of Marine Engineering, AMET University, Chennai, Tamilnadu, India

2 Department of Mechanical Engineering, AMET University, Chennai, Tamilnadu, India

3 Department of Mechanical Engineering, K. Ramakrishnan College of Technology, Trichy, Tamilnadu, India

*Address all correspondence to: praveen71989@gmail.com

References

[1] Panwar NL. Design and performance evaluation of energy efficient biomass gasifier based cookstove on multi fuels. Mitigation and Adaptation Strategies for Global Change. 2009;**14**(7):627-633

[2] Pathak BS, Patel SR, Bhave AG, Bhoi PR, Sharma AM, Shah NP. Performance evaluation of an agricultural residue-based modular throat-type down-draft gasifier for thermal application. Biomass and Bioenergy. 2008;**32**(1):72-77

[3] Nsamba HK, Hale SE, Cornelissen G, Bachmann RT. Sustainable technologies for small-scale biochar production—A review. Journal of Sustainable Bioenergy Systems. 2015;**5**(01):10

[4] Boob TN. Performance of saw-dust in low cost sandcrete blocks. American Journal of Engineering Research. 2014;**3**(4):197-206

[5] Jain M, Mudhoo A, Garg VK. Swiss blue dye sequestration by adsorption using Acacia nilotica sawdust. International Journal of Environmental Technology and Management. 2011;**14**(1-4):220-237